ISBN 978-1-331-95507-8
PIBN 10259258

English
Français
Deutsche
Italiano
Español
Português

www.forgottenbooks.com

Mythology Photography **Fiction**
Fishing Christianity **Art** Cooking
Essays Buddhism Freemasonry
Medicine **Biology** Music **Ancient
Egypt** Evolution Carpentry Physics
Dance Geology **Mathematics** Fitness
Shakespeare **Folklore** Yoga Marketing
Confidence Immortality Biographies
Poetry **Psychology** Witchcraft
Electronics Chemistry History **Law**
Accounting **Philosophy** Anthropology
Alchemy Drama Quantum Mechanics
Atheism Sexual Health **Ancient History**
Entrepreneurship Languages Sport
Paleontology Needlework Islam
Metaphysics Investment Archaeology
Parenting Statistics Criminology
Motivational

THE ESTABLISHMENT OF VARIETIES IN COLEUS BY THE SELECTION OF SOMATIC VARIATIONS.

BY

A. B. STOUT,

Director of the Laboratories, New York Botanical Garden.

WASHINGTON, D. C.
PUBLISHED BY THE CARNEGIE INSTITUTION OF WASHINGTON
1915

Carnegie Institution of Washington
Publication No. 218

4 4 0 9

Copies of this Book
were first issued
OCT 7 1915

PRESS OF GIBSON BROTHERS, INC.
WASHINGTON, D. C.

TABLE OF CONTENTS.

THE ESTABLISHMENT OF VARIETIES IN COLEUS BY THE SELECTION OF SOMATIC VARIATIONS.

By A. B. Stout.

INTRODUCTION AND HISTORICAL REVIEW OF LITERATURE.

Judgment as to the genetic constitution of a plant is based on the expression of characters in a plant itself, in members of a selfed progeny and in a hybrid progeny. It is largely through the study of the last named that there has developed the conception that characters are represented in cells by unit factors. In considering the evidence as to whether these assumed factors are in any sense units, it is highly essential that the individual be studied as to the variability of characters and the range of expressions exhibited in homologous parts. For this purpose so-called bud variations are especially significant, since they represent, perhaps, the extreme of spontaneous somatic variability and suggest that quite as marked variations as exist in hybrid progeny may develop in the parts of a single individual.

In an excellent summary of all cases of bud variation known to him, Darwin (1868) shows that the phenomenon is widely distributed in the plant kingdom and that it may affect various parts of a plant. He drew the highly interesting conclusions that they include: (1) reversions to characters not acquired by crossing, (2) reversions in hybrids to parental characters, and (3) cases of spontaneous variability. The latter he considered as of the same sort as appear in seed progeny. According to Darwin's views, "long-continued and high cultivation" are conditions that induce bud variations, but he recognized that certain cases, especially those when only single buds or parts of buds are changed, do not seem to be due to external conditions. Darwin considered in general that bud variations are evidences of the almost unlimited variability that plants manifest, due to the nature of living structures and the exciting causes of environment.

Darwin did not believe in fixed hereditary units. He held that characters may respond directly to the effects of environment, and also exhibit spontaneous variability both in seed and in vegetative propagation. Furthermore, sexual hybridization was considered to be of influence in modifying and changing inherited characteristics.

De Vries (1901, vol. I, p. 39) considers that bud variations are spontaneous changes most common in varieties with incompletely fixed characters. He includes (1889, p. 13) these phenomena under the head of "dichogeny," a general conception proposed for cases in which the nature of the organ may be determined, as he assumes, by external

3

conditions, but he also assumes that the internal constitution may admit of development in several directions.

De Vries discusses bud variation with special reference to variegation. For the frequent and striking cases of the production of variegated branches on green plants and the development of green branches on variegated plants, he offers the old explanation of "latent potentiality." They belong in general with mutations in that they appear as clear-cut discontinuous variations; the former are classed as progressive mutations and the latter retrogressive (1901, vol. I, p. 606).

De Vries makes a most comprehensive analysis of the nature of variegation and concludes (1901, vol. I, p. 616) that the capacity for variegation is more widely distributed in the plant kingdom as a latent or semilatent character than perhaps any other character. He notes that true *aurea* varieties are few and are remarkably constant. Most variegated races show rather wide fluctuations and constitute what he calls intermediate races. His scheme (1901, vol I, p. 424) of representing the relationship is as follows:

	Normal character.	Anomaly.
I. Original species, green..	Active.	Latent.
II. Half race, rarely variegated	Active.	Semilatent.
III. An equilibrium is maintained.		
IV. Eversporting variety, variegated....	Semilatent.	Active.
V. Constant variety, *aurea*...	Latent.	Active.

In a later publication (1913) de Vries considers that the pangens, which he assumes are directly concerned with the transmission and expression of characters, may be not only active, semilatent, and latent, but also labile.

The transition from constant green varieties through variegated varieties to constant *aurea* varieties is conceived to be dependent on the degree of activity displayed by antagonistic qualities within the cells. The essential changes are conceived to be intracellular and not dependent on qualitative cell-divisions; all the cells are potentially alike, but different processes within the cells give differences in expression. Thus the conception of de Vries does not consider that the hereditary physiological units (pangens) are fixed and uniform. They are subject to effects of environment and they exhibit spontaneous change even to the degree of sudden appearance by progressive mutation. All these phenomena may be exhibited either in vegetative or in seed progeny. Sexual hybridization may entirely modify or change the effect and nature of the basic physiological units of heredity. De Vries's general attitude, however, places the emphasis on discontinuous or mutational changes as the only really stable variations.

The exhaustive summary of facts given by Cramer (1907) fully substantiates the views of Darwin and de Vries as to the main classes of variegation and the remarkable gradation in degrees of constancy and inconstancy which are exhibited by the various types and by their bud variations. Cramer gives the most complete and detailed summary of the known cases of bud variation that has been published. He includes (p. 18) under the term "Knospenvariation" all cases where a character suddenly changes in a plant in a way that can not be attributed to environmental influence. He considers that there are three main classes of bud variation: (1) vegetative segregations in hybrids; (2) intermediate-race bud variations; and (3) vegetative mutations. Recessive characters in hybrids may, he assumes, separate out by vegetative cell-divisions, and dominant characters which have been latent may reappear. His conception of the variability of characters in an intermediate race is the same as that of de Vries. Characters concerned in bud variations in intermediate races show, it is considered, great irregularity in expression. The bud variations which give a seed progeny quite constant for the character involved are classed as vegetative mutations. This classification emphasizes the fact that characters involved in hybridization may come into expression quite irregularly, and that spontaneous and fluctuating variations are common in vegetative development.

Cramer devotes a most interesting and instructive chapter to variegated plants, showing, especially, their wide distribution in the plant kingdom, the wide fluctuation in the degree of constancy of their seed progeny, the range from vegetatively constant to inconstant types, and the frequency of bud variations in variegated plants. He considers (pp. 126, 127) that loss of variegation can occur in two ways: (1) by atavistic bud variations and (2) through influence of external conditions. He notes that it is very difficult to distinguish fluctuating variability from mutation.

The general variability of the characters concerned with variegation is well shown in his discussion of changes that may occur in a single plant, such as the following: In apparently inherited types of variegation, the seedlings are often green at first. Seedlings may have at first variegated cotyledons, then a few green leaves, and then variegated leaves. Some biennial plants are pure green during the first year of growth, but variegated in the second year. Various plants exhibit a periodicity in their variegation, being green in spring and variegated later in the summer, or *vice versa*. He gives numerous cases of the appearance of variegation by bud variation on green plants resulting in new varieties or in the duplication of types already known. These exhibit various degrees of constancy.

Various adherents of Mendelian doctrines have more recently discussed the transmission of characters appearing as bud variations, attempting to explain them, as they do all heritable variation, by the presence or absence of a unit factor. While this is the general attitude of those who have investigated the inheritance of variegation and the nature of bud variations involving variegation, the results of their studies, as a whole, show great diversity and many necessary modifications of the general Mendelian doctrine of the integrity of unit characters or unit factors and the purity of the segregations of such assumed factors.

East (1908, 1910 a) considers that a large majority of the known cases of bud variation are due to the loss of a dominant character and that 70 per cent of all known cases are color variations. His discussion and suggestions do not claim to be comprehensive or critical, and he excludes from his treatment all cases of bud variations in variegated plants, because some types of these are known to be pathological. It is interesting to note that he states that no important potato has arisen as a bud sport. He reports four cases of bud variation in potatoes giving white from red or pink which appear to be constant; also several cases giving colored or purple blotched from pure white, concerning which no data are given except the statement that they are not constant. East's view is that bud variations are due to loss or latency of hereditary units that stand for characters.

Bateson (1909, p. 273), especially, has advocated, on theoretical grounds largely, that bud variations are due to qualitative cell-divisions in somatic tissues, giving somatic segregation of unit factors. The idea is quite identical in its main features with Weismann's conception of qualitative divisions, giving tissue differentiation in ontogeny.

In considering the nature of the albomarginate types of variegation, with special reference to the origin and the development of the green and white areas, Baur (1909) has made criticial anatomical studies of a white-margined *Pelargonium zonale* which indicate very clearly the relationship of white and green tissues as periclinal chimeras and explain the appearance on them of branches wholly green or white. In testing, by crossing experiments, his assumption that the green and white tissues are each pure, he is forced to the further assumptions that the loss of green in this case is due solely to the condition of the chromatophores and that male sex-cells carry chromatophores. He does not consider that white cells can arise spontaneously from green, or *vice versa*. Yet this probably did occur when the chimera was first produced. In fact, the numerous varieties that are peripheral chimeras not only in *Pelargonium* but also in other genera indicate that such spontaneous loss of power to produce green is frequent. We may say that Baur's interesting results, however, do show that if spontaneous loss occurs in young leaves after they are formed, mottled or striped variegation is produced, but if the loss occurs in the growing-point itself, then chimeras will result, their constancy depending largely on the relative permanency of the change.

Later, Baur (1910), in studies of seed progeny of variegated types that appeared spontaneously, assumes that green is the combined result of three factors. If a certain one (Z) is absent the tissue is colorless, if Z and Y only are present then a *chlorina* type is produced, and if N is present with Z and heterozygous Y, the *aurea* coloration results. In *Antirrhinum majus albomaculatum*, Baur found 8 types of variegation arising spontaneously in cultures of about 30,000 pure-green plants. His crosses with these show that certain cases appear to be inherited only from the seed parent. To explain the results, Baur assumes that hereditary qualities are localized in different parts of the cell. The nucleus, the cytoplasm, and the chromatophores all possess, he considers, different but definite factors concerned with variegation.

Such results and conclusions illustrate very well the difficulties and uncertainties which arise from attempts to analyze variations in terms of unit factors and suggest most forcibly the need of a more thorough investigation of such variations in a progeny derived by vegetative propagation.

The variations among the branches of a single plant which *Correns* (1909, *a* and *b*) reports in connection with variegated types of *Mirabilis jalapa* are apparently quite similar, in degree at least, to those I shall report for *Coleus*. It would seem to be highly important that the inheritance of these variations be studied in vegetative propagation. *Correns*, however, made a study of seed progenies only. In the case of the *albomaculata* type these were composed of green, white, and *albomaculata* plants in quite different ratios for different plants tested. All these classes appeared when pollen from a pure-green plant was used, but when pollen from the *albomaculata* was used on pure green there was no transmission of the quality of variegation. This case of matrocliny is due, he assumes, first, to the localization in the cytoplasm of the factor for variegation, and second, to the condition that male sex-cells in this case do not carry cytoplasm.

Shull's (1914) studies with variegated types of *Melandrium* show much the same results as those of Baur and *Correns*. He distinguishes between *chlorina, pallida*, and pure-green types of *Melandrium* on the basis of presence or absence of three factors which in crosses behaved quite like units. In types with *green-white blotched* and with *chlorinomaculata* variegation, however, the variegation seemed to be transmitted only through the seed parent, but not uniformly, for crosses of variegated branches with pure green gave in the F_1 generation plants ranging from pure green through types of *chlorinomaculata* to yellowish-green plants. It is highly interesting that Shull found that the greater the amount of *chlorina* coloration in the calyx the greater was the number of variegated seed progeny. The *aurea* types, which possessed as a rule small round flecks, gave such varied results with appearance of different types in F_1 progeny that Shull concludes it must be an infectious variegation which is transmitted through both germ-cells to a part of the progeny.

Among the most interesting series of observations especially bearing on the behavior of red coloration in *Coleus* and most important in the consideration of the nature of variegation and the character of somatic variation and the relations of these to seed progeny, is that of Emerson (1914). The variegation in question is that of pericarp color in certain "calico" races of corn. The extent of coloration varies widely, ranging from solid red through every degree of striping and blotching to entirely non-red, both for ears as a whole and for single kernels on the same ear.

Emerson made careful studies of the progeny of seeds having different degrees of coloration. His results show a wide range of variation in the progeny of kernels that appear to be identical. Selected solid-red kernels from "freak ears" of unknown parentage gave, in some cases, progeny with only solid-red or non-red ears, and in other cases the plants produced solid-red, variegated, or non-red ears. Variegated and white kernels (data not given separately) gave either variegated and white, or red, variegated, and white. Again, from two ears, kernels of white gave progeny pure white, and red kernels gave red and white only, each of which gave afterward a constant progeny.

In selfed variegated strains, kernels of all classes from solid-red to non-red gave progeny with ears solid-red or variegated, but none with no-red ears.

Data are given collectively for progeny of 5 solid-red ears (selfed). These gave solid-red and variegated (p. 18). *Progenies of only two plants of these solid-reds are reported.* One gave again plants with solid-red or variegated ears, the other gave only solid red. The numbers grown in this generation were respectively 9 and 16. Emerson considers from these data that, in general, red-eared plants behave, judged by progeny, as if they were hybrids either between solid-red and variegated or solid-red and white races.

The data show quite clearly, as Emerson points out, that the more red there is in the seed planted the larger the percentage of red ears in the progeny. The variegated race is therefore far from constant. Selection from red kernels and from red ears give a strain quite constant for solid-red, but Emerson's data are far from conclusive that a pure solid-red strain was obtained in this way.

The results of crossing the variegated race with non-red races are interesting. When the former was the pollen parent, the F_1 progeny gave red, variegated, and non-red ears, but Emerson states that some of the latter may have been extremely light types of variegation. From the reciprocal cross selected kernels for the F_1 gave red and variegated or only variegated.

In the selections from hybrid stock, as in that in selfed stock, the seed that had more red gave greater numbers of pure-red progeny.

Although there is this irregular expression of solid-red, variegated, and non-red kernels on the same plant and among different plants of the same generation, and although selection from solid-red tends to give greater numbers of solid-red progeny, Emerson concludes that the factors for solid-red and variegation are "as distinct in inheritance as any two factors could well be" (p. 33). He points out, however, that the "factors" concerned are pattern factors, one determining self-color and the other giving variegation. When a solid-red kernel occurs in an ear of the variegated race, Emerson assumes that a V factor changes to $S;$ but when these red kernels do not give solid-red progeny, he further assumes that in these diploid cells one of the pair of V factors changes to S and that such somatic changes may affect an area of cells including macrospores.

Taking this at its full value, we note that as the occurrence of solid-red kernels is frequent the hereditary factor V is fluctuating and extremely labile, changing to S readily. For V and S to be distinct in inheritance under such conditions is hardly conceivable, for they are not even distinct in cell lineage. Emerson thus reflects the strong tendency of most modern Mendelian investigators to regard their assumed factors as temporary conditions with quite fluctuating activities.

Such conflicting results as the above, obtained even by careful pedigree methods of study, may well lead one to question whether our knowledge of the behavior of plant characters in inheritance and expression has advanced much beyond the views of Darwin.

Such studies may well establish the general breeding values of certain characters in particular cases which are of practical significance. The more refined methods of pedigree have shown that plants with identical appearances may give quite different progeny, and that selection for relative degrees of constancy should be made in pedigreed lines rather than in mass selection. Mendelian results also indicate the possibilities of hybridization followed by selection in pedigreed lines of hrybrid progeny.

In its theoretical significance, however, Mendel's original work has two points of special interest. First, it embodied the conception that all structures of like character are due to a single hereditary unit; for a specific example, all the wrinkles of all the peas on a plant were conceived as represented in the germ-cells by a single unit. This was in decided contrast to Weismann's general view that each wrinkle is represented in the germ-cells. The second point of special interest pertains to the behavior of these assumed units in the formation of reproductive cells and their behavior in fertilization. Mendel assumed that they segregated as units independent in behavior and pure in composition from the unit representing the contrasting character. By the conception of the purity of the segregations of hereditary units, each representing only an entire and complete quality or character, Mendelism must be judged.   It is not a question of the appearance of parental characters in hybrid progeny, but of the purity of those segregations.

Mendel himself greatly modified the conception, gained from his studies with peas, that an entire quality is represented by a single unit.   In *Phaseolus* he crossed *P. nanus* having white flowers with *P. multiflorus* having purple flowers.   The F_2 generation of 31 plants had flowers ranging from white to pale violet and purple red of various grades.   Only one had white flowers like those of *P. nanus*.   He suggests that the color of flowers (and seeds as well) in *P. multiflorus* "is a combination of two or more entirely independent colors" (Mendel, p. 367).   To Mendel, therefore, should be given credit for the conception of multiple factors, later developed especially by Nilsson-Ehle (1909) and by East (1910 b).

Very soon after the so-called rediscovery of Mendel's law for the behavior of certain characters of *Pisum* in hybridization, it was noted that new qualities frequently appear in the F_2 generation, as Mendel found was the case in beans.   The presence-and-absence theory developed by Bateson and Punnett (1905) attempted to account for this increased variability giving the appearance of new types.   This gave chance for dihybrid ratios from what was apparently a simple pair of contrasting characters, the "absences" segregating out into expressed characters new at least to the immediate ancestry.

The next important modification of the general view was the application of Mendel's idea of what we now call multiple factors.   Nilsson-Ehle (1908 and 1909) found that certain crosses in cereals gave in the F_2 progeny large numbers of one parental type and relatively few of the other type.   For example, the apparently simple cross of white-kerneled wheat (Predel) with a red-kerneled sort (Swedish Velvet Chaff) gave an F_2 progeny of about 63 red-kerneled plants to 1 white.   Nilsson-Ehle's explanation assumes that the red character in the Swedish Velvet Chaff is due to three independent factors which are each of equal value and that any one can produce the same effect as all three.   We note that the variability of the F_2 generation was not increased over that of parents, but that the ratio showed almost complete appearance of one character.

East (1910 b) applied the term to the same sort of phenomena as Mendel did, *i. e.*, to increased variability.   The apparently simple cross of white with yellow corn gave in the F_2 a generation with few white but with a large number of yellow kernels.   The latter, which were of all gradations of intensity, were grouped into two classes by East.

It should be noted that such a group of "factors" having among themselves different values but working together to produce a single character are not necessarily independent in the fertilizations of the variety concerned.   Inside the variety they go together.   When certain crosses are made they separate.   This is but another way of saying that the hereditary bearers of characters appear to be split up and modified by crossing, giving in some cases quite new characters.

Thus in the period of less than 20 years the literature passes from the confident treatment of *characters* as units with a simple shorthand representation to a discussion of *"factors"* that under some conditions work together as a multiple unit and under other conditions separate, producing equal or different expressions. The assignment of different values to the assumed factors as diluters, intensifiers, inhibitors, and the conception of multiple factors that can separate out, giving aberrant ratios or new expressions and even almost endless intermediate gradations, both of so-called qualitative and quantitative characters all reduced to descriptive terms, add nothing to the fundamental conceptions of Darwin regarding variability in hybridization. The extreme application of the multiple-factor hypothesis simply means that small variations are inherited equally as well as are large variations.

The greater number of Mendelians, mutationalists, and adherents of the doctrines of pure lines seem to hold that the unit factors are changeless. Others still accept Darwin's general views of the modifiability of the fundamental units. The latter view is especially well developed by Castle (1912), who states (p. 356):

"In my experience every unit-character is subject to quantitative variation, that is, its expression in the body varies, and it is clear that these variations have a germinal basis because they are inherited."

Morgan (1913) considers that factors are labile aggregates subject to rearrangement, that processes of mutation and reversion are reversible, and that in eversporting varieties mutation and reversion are regular processes.

Bateson (1902, p. 201), in a defense of early Mendelian views, makes the following admission:

"We have to consider the question whether the purity of the gametes in respect to one or another antagonistic character is or is likely to be in the case of any given character a universal truth. The answer is unquestionably *No*, but for reasons in which ancestry plays no part."

More recently (1914, p. 322) he has expressed the view that the conception of multiple factors is in his mind an admission that there are imperfect segregations. To quote further:

"Segregation is somehow effected by the rhythms of cell-division, if such an expression be permitted. In some cases the whole factor is so easily separated that it is swept out at once; in others it is so intermingled that gametes of all degrees of purity may result."

Thus it seems that the short-hand system of representing hypothetical germ-cell units is often not only cumbersome but inaccurate. Some phases of it may be useful as descriptive terms, but the method has led into purely speculative fields in the attempts to represent and explain imperfect segregations and variations in the hereditary qualities.

Furthermore, it appears in final analysis that the extended studies of seed progenies have not contributed anything fundamentally new to the knowledge of the nature of plant characters. At least, the analysis of characters in terms of hereditary units has failed.

It would seem that, in considering these problems, the studies of variation among numbers of a seed progeny is no more important than studies of variation in progenies derived by vegetative propagation. The latter should give much more conclusive data regarding such questions as the constancy of characters (or of assumed factors), the purity of apparent segregations, and the frequency, constancy, and nature of spontaneous changes in the expression of characters.

It is evident that the facts regarding bud variation involve the fundamental questions of heredity. When such variations occur in a plant that can be propagated vegetatively, there is opportunity to apply the pedigree method of experimental study to successive generations produced by vegetation propagation. The nature, frequency, and permanence of such changes as appear can be studied without the complications that are associated with alternation of generations and fertilization as they normally occur even in selfed seed progeny. Special evidence regarding the "expression" of characters, which also bears on the question of their inheritance, may thus be obtained.

THE PROBLEM IN COLEUS.

For the study of variation along the lines indicated above, I have grown a series of 833 plants, all descended by vegetative propagation from two plants of a variety of *Coleus*.

Coleus is particularly favorable for such study in that bud variations are frequent and the plant is readily propagated vegetatively. The leaves are in pairs which alternate on a square stem, making but four rows of leaves. Bud variations that appear sectorially can thus be traced with ease. In a young plant lateral branches usually start to develop from the axils of all the leaves on the main stem. In a large, bushy plant many lateral buds remain dormant, but by proper pruning any bud can be forced to develop or it can be propagated as a cutting which will give it full chance for development. Large, bushy plants 3 or 4 feet tall grown out-of-doors often have a total of as many as 300 branches. In the greenhouse, also, plants can be grown from cuttings, with the production of many leaves and branches. The ease with which the plant is propagated vegetatively makes it possible to grow a large series of pedigreed plants from any cutting, and to thus test the frequency with which variations appear, the constancy of the different types, and the purity of any vegetative segregations, all bearing on an analysis of the nature and inheritance of the characters concerned.

METHOD OF RECORDING RESULTS.

To record fully the series of *Coleus* plants descended through vegetative propagation from any plant or any particular branch used as a cutting, the pedigree method of culture has been used.   It may be mentioned that the application of the pedigree-culture method to plants propagated vegetatively is much simpler than its use in seed progenies, the difficulties of which have been ably presented by Shull (1908).

Each cutting was given a number which indicated its lineage.   The two plants serving as original parents were numbered 1 and 3, and the first digit to the left in a number of any plant indicates the original parent plant.   For the progeny of plant No. 1, the second digit indicates the particular main lateral branch from which the first cuttings were taken, and the third digit is the number of the particular cutting. Additional digits indicate successive plants grown later in the particular line of descent.   Thus, plant 121 (I read these one-two-one, etc.), 122, and 123 were grown from three cuttings taken from branch No. 2 of plant No. 1.   Plants 1211, 1212, and 1213 were from cuttings of plant 121, while the numbers 12111, 12121, 12131, etc., were given to plants of the next generation of cuttings.   Thus the number of any plant gives its complete pedigree since the experiments were begun.   Particular data regarding the cuttings and the plants have been recorded on cards and filed in the manner of a card catalogue.   This enables one to trace readily the inheritance of any variation through a series of generations and to compare different lines of descent from any point in the culture.

Following the suggestions of Webber (1903) and Shull (1912) the term "clone" will be used in speaking collectively of all plants descended from any one plant or branch.   All the plants derived from plant No. 1 constitute a main clone, itself made up of numerous subclones.   The records of pedigree enable me to designate these as clone 11, clone 13, clone 117, etc.   I shall use the term "line of descent" to include the different plants that were the parents of any one plant.   The term "generation" refers to the plants that were grown during the same period.

The observations here reported were made on successive generations of pedigreed plants derived by vegetative propagation from two original plants.   These parent plants were alike when young in possessing a color pattern that can be characterized as a mosaic of green, yellow, and red, with the colors distributed as shown in figure 2.

In referring to the color patterns, it seems best to the writer to use terms that are sufficiently descriptive to make the matter concrete and which at the same time are somewhat compact.   The color of the subepidermal tissues in the center of the leaf will be mentioned first, as green or yellow; the color of the subepidermal tissues at the margin will be mentioned next, as green or yellow, and last the character of the epidermis as *red blotched* (both surfaces), *solid red* (both surfaces), or *solid red upper center*. In these compound expressions, hyphens will be used to separate the terms descriptive of each color element. In the cases with colorless epidermis no reference is made to this condition. The colors were determined by Ridgway's Color Standards and Color Nomenclature. The color pattern of the parent plants, in this paper referred to as a *green-yellow-red blotched* pattern, is a mosaic made up of a green center and a yellow border with conspicuous epidermal blotches of red. The yellow is *amber yellow* and constitutes an irregular band about the margin of the leaf. The green is a *spinach green* and is chiefly massed in the center of the leaf. Over the green portions the red appears as *violet carmine*, but over the underlying yellow areas it is *nopal red*. The three color elements are in such sharp contrast that any marked variation is readily noted. Increase or loss of either the yellow, green, or red is conspicuous, as one will appreciate from a glance at the plates that illustrate this article.

Spontaneous bud variations consisting of marked alterations in color pattern appear either in single leaves or groups of leaves, or in single branches or groups of branches, affecting the whole or a part of the leaves or branches. When appearing in a terminal bud, one or more leaves have a pattern differing from that of the leaves below. When appearing in a lateral bud, the first leaves of the branch possess a pattern different from that of the subtending leaf. Those appearing in a terminal bud have, in all cases observed by the writer, been sectorial in the main branch itself. That is, the change has appeared first in a part of the branch only. These variations carried on into newly formed branches give plants bearing two, three, or even four distinct types of foliage, with differences especially marked in cases of single branches with sectorial distribution of two patterns. The rather simple arrangement of the leaves and branches in *Coleus* enables one to trace the extent of a variation.

This may be illustrated by a variation that occurred in plant No. 1171. In this variation the relative positions of the green and the yellow became reversed, as shown in the two leaves reproduced in figures 2 and 6. When the cutting was made in April 1912, all the leaves had uniformly green centers. On one of the first pair of branches to develop, however, all of the leaves had the yellow in the center. As further branches developed, the new pattern appeared in 5 other branches. The plant produced 13 pairs of branches on the main branch before it was necessary to remove the terminal bud to insure proper development of the lateral branches. All of these branches developed at least to a size sufficient to show the color pattern of the leaves borne. The 6 branches with the new pattern were contiguous and were located as indicated in diagram 1. The plant was transplanted to a large pot and kept in a greenhouse over winter and then grown out of doors during the summer of 1913. All of the leaves produced by the 6 branches in question in the 17 months of growth were uniform and constant to the new type and were in marked contrast to the foliage of the other branches. During the summer of 1913, two bud variations occurred in secondary branches in the upper part of the plant. One was a sectorial loss of yellow giving type *green-red blotched* from *green-yellow-red blotched* and the other was a complete loss of green in one branch giving type *yellow-red blotched*. In September 1913, the plant bore four distinct kinds of color pattern, viz, *yellow-red blotched* (fig. 1), *green-yellow red blotched* (fig. 2), *green-red blotched* (fig. 5), and *yellow-green-red blotched* (fig. 6). The bud variation to type *yellow-green-red blotched* was sectorial in the main stem for a vertical distance of six nodes, but was not complete for the entire stem, a condition shown in diagram 1.

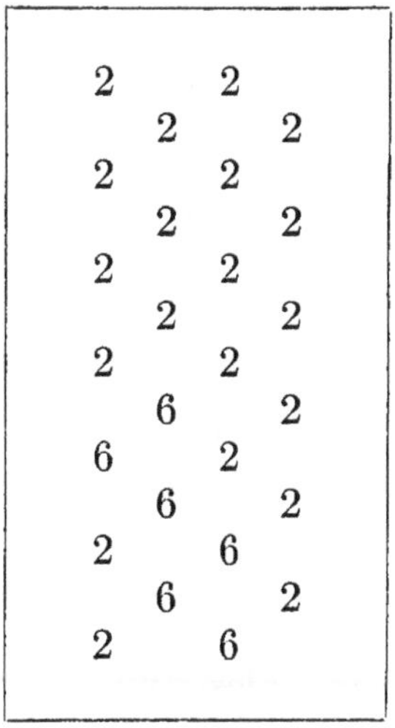

The greater number of bud variations first appeared in single lateral branches and not in a series of branches on a main stem, as described above for 1171. Where such a variation was sectorial in a branch the continued growth gave more or less irregular extension of the new type.

The parent plant here designated as No. 1 was one of several *Coleus* plants which were grown at the New York Botanical Garden during the summer of 1911. This plant possessed in September 1911, when first observed by the writer, two branches bearing leaves in which the yellow was apparently almost entirely absent. These two branches were in the same rank, one directly above the other. About one-third of the entire foliage of the plant was borne by these two branches and the marked green aspect of this part of the plant was in decided contrast to the conspicuous yellow in the foliage of the rest of the plant. Upon careful examination, a few yellow spots could be seen in many of the leaves of one of the green branches (branch 14) quite like those of the leaf shown in figure 4. The leaves on the other green branch (branch 13) were apparently free from all yellow areas (fig. 5).

The decided loss of yellow in these two branches constituted the only variation in the dozen or more plants in this particular bed of *Coleus*. To test the constancy of this variation, as well as the reappearance of it and of other variations, pedigreed cuttings were made from each of four main lateral branches of the plant.

About the same time random cuttings were made from the bed of plants for stock for general planting. One of these cuttings produced

DIAGRAM 1.—Position of the six branches on plant 1171 having color-pattern *yellow-green-red blotched* (6) among those having *green-yellow-red blotched* patterns (2).

during the winter of 1911–12 a single branch bearing leaves with the red completely covering both surfaces. This color pattern designated as *green-yellow-solid red* is shown in figure 8. All the leaves on the branch were uniform for this pattern and were in most conspicuous contrast to the rest of the plant. From this plant a number of pedigreed cuttings were also made.

METHOD OF HANDLING CULTURES.

The first generation of plants (series 111, 121, 131, 141) was grown in a greenhouse during the winter of 1911–12. In April 1912, cuttings (series 1111, 1211, etc.) were made from these. During the summer all of the plants, both old and young, were grown out of doors in beds. In the autumn cuttings were again made. The plants developed from cuttings taken in the autumn were under observation for a year, 7 months of which they were grown under greenhouse conditions. Cuttings taken in spring were grown only out of doors. This method of handling gave opportunity to observe development and behavior under different conditions and to compare old plants with younger ones. Except for a few plants that were subjected to special conditions, all the plants of any generation were treated uniformly with respect to kind of soil, size of pots, and conditions of temperature and illumination. The plants were cut back somewhat to prevent early blossoming and to maintain a vigorous vegetative condition.

In the period of three years between September 1911 and September 1914, a total of 833 plants were grown to maturity and discarded. All of these descended through vegetative propagation from plants Nos. 1 and 3, both of which had originally the *green-yellow-red blotched* pattern illustrated by figure 2.

GENERAL SURVEY OF THE VARIATIONS.

Variations in the color patterns of the plants both of the original and the derived types can be classed as fluctuating variations and as bud variations. In the former the changes were usually quite gradual and affected in most cases an entire plant. The changes which are in this paper included in the term "bud variations" were those affecting only a part of a plant and usually appearing as a sudden and conspicuous change. In addition to the variations in color patterns, there appeared, in several subclones, plants which fluctuated in leaf-shape, giving in extreme cases leaves deeply cut and laciniate. The variations that appeared were as follows: (*A*) changes involving yellow and green; (*B*) changes involving the epidermal red, and (*C*) changes involving leaf-shape. The bud variations can be grouped as in table 1 with data regarding the number of plants concerned and the number of times the different changes appeared as a bud variation.

TABLE 1.—*Types and frequency of the bud variations.*

Types of bud variation.	Number of plants concerned.	Frequency of the variation.
A. Changes involving yellow and green:		
I. Increase of yellow and decrease of green:		
1. *Yellow-red blotched* (fig. 1) from *green-yellow-red blotched* (fig. 2)	337	7
Yellow-red blotched (fig. 1) from *green-yellow spotted-red blotched* (fig. 4)..	198	1
Yellow-red blotched (fig. 1) from *yellow-green-red blotched* (fig. 6)..........	41	3
Yellow-solid red from *green-yellow-solid red* (fig. 8)..	54	1
2. *Green-yellow-red blotched* (fig. 2) from *green-yellow spotted-red blotched* (fig. 4)	198	3
Green-yellow-red blotched (fig. 2) from *laciniate, green-yellow spotted-red blotched*	68	1
3. *Green-yellow spotted-red blotched* (fig. 4) from *green-red blotched* (fig. 5)....	90	2
4. *Spontaneous yellow* from *green-red blotched* (fig. 5)......................	90	9
II. Decrease of yellow and increase of green:		
1. *Green-yellow spotted-red blotched* (fig. 4) from *green-yellow-red blotched* (fig. 2).	337	8
Green-yellow spotted-red blotched (fig. 4) from *yellow-green-red blotched* (fig. 6).	41	1
2. *Green-red blotched* (fig. 5) from *green-yellow-red blotched* (fig. 2)	337	24
Green-red blotched (fig. 5) from *green-yellow spotted-red blotched* (fig. 4)....	198	2
Green-red blotched (fig. 5) from *laciniate, green-yellow spotted-red blotched*...	68	1
Green-red blotched (fig. 5) from *yellow-green-red blotched* (fig. 6)..........	41	10
Green-solid red (fig. 9) from *green-yellow-solid red* (fig. 8)................	54	4
III. Reversal of position of green and yellow:		
1. From yellow border to yellow center		
a. *Yellow-green-red blotched* (fig. 6) from *green-yellow-red blotched* (fig. 2).	337	6
b. *Yellow-green-solid red* (fig. 11) from *green-yellow-solid red* (fig. 8) ...	54	2
B. Changes involving the epidermal red:		
I. Increase of red:		
1. *Green-yellow-solid red* (fig. 8) from *green-yellow-red blotched* (fig. 2).......	337	4
2. *Green-yellow spotted-solid red* from *green-yellow spotted-red blotched* (fig. 4).	198	2
3. *Green-solid red* (fig. 9) from *green-red blotched* (fig. 5)	90	2
II. Decrease of red:		
1. *Green-yellow spotted* from *laciniate, green-yellow spotted-red blotched*........	68	1
2. *Green-yellow spotted* from *green-yellow spotted-red blotched* (fig. 4) ...	198	2
3. *Green-yellow* (fig. 12) from *green-yellow-red blotched* (fig. 2)..............	337	10
Green-yellow (fig. 12) from *green-yellow-solid red* (fig. 8)..	54	1
4. *Green* (fig. 13) from *green-red blotched* (fig. 5)......................	90	3
Green (fig. 13) from *green-solid red* (fig. 9).........................	8	1
5. *Yellow-green* (fig. 14) from *yellow-green-red blotched* (fig. 6).............	41	1
III. Decrease of red with concentration in epidermis of upper surface:		
1. *Green-yellow-solid red upper center* (fig. 10) from *green-yellow-solid red*....	54	2
C. Changes involving leaf-shape:		
1. From *entire to periodically laciniate*..	765	13
2. From *periodically laciniate to constantly entire*........................	68	1

The names given to the different patterns embody the principal features of coloration on the basis explained above (see pp. 13 and 14). The patterns selected are with one exception those that appeared as conspicuous bud variations and which are sufficiently distinct for ready identification. Numerous other types that are intermediate between the types given could also be designated by still more exact classification. The following descriptions, together with the colored plates illustrating the types as classified, will enable the reader to visualize the patterns referred to by name.

Color pattern *yellow-red blotched* (fig. 1): Leaves almost entirely *amber yellow* with only very limited and scattered areas of greenish tissue. The island-like areas of green are surrounded by yellow. Irregular-shaped blotches of *nopal red* are scattered over both upper and lower surfaces. This decidedly *yellow* pattern was derived from the several patterns, as shown in table 1, by a sudden and a conspicuous loss of green tissue.

Color pattern *green-yellow-red blotched* (fig. 2): This is the pattern possessed originally by the two parents and has already been described.

Color pattern *green-yellow spotted-red blotched* (fig. 4): In this pattern there is *no definite border of yellow*. The yellow appears in rather limited and somewhat scattered areas, sometimes nearly limited to the border zone, but often quite generally distributed throughout the leaf. The pattern is, therefore, decidedly greener in appearance than that of the parental type.

Color pattern *green-red blotched* (fig. 5): This is a bicolored pattern of green and red. As there is no underlying yellow the epidermal red appears uniformly as violet carmine. This type arose frequently on plants with patterns containing yellow by what was apparently a complete loss of yellow.

It may be noted that in the four patterns as arranged above there is an increase of green and a corresponding decrease of yellow, with the distribution of the epidermal red quite uniform. The *yellow-red blotched* pattern gives the extreme development of yellow with almost complete absence of green. The *green-red blotched* pattern has apparently a complete loss of yellow. The *green-yellow-red blotched* and the *green-yellow spotted-red blotched* patterns are gradations between these extremes.

Color pattern *yellow-green-red blotched* (fig. 6): This is a pattern of green, yellow, and red as in type *green-yellow-red blotched,* but the relative positions of the green and the yellow are reversed. The yellow is in the central portion of the leaf.

Color pattern *green-yellow-solid red* (fig. 8): Both surfaces of the leaf are a solid red. Through the center of the leaf the color is *violet carmine,* but the marginal zone underlaid by yellow is *nopal red.* At the base of the leaves a greenish tint prevails and at the extreme edge of the margin a fine line of yellow is visible. On the under surface the red seems slightly less intense and does not cover the larger veins, which stand out prominently on this surface. This pattern differs from *green-yellow-red blotched* in having the entire epidermis solid red instead of blotched. Frequently, however, a few isolated areas are free of epidermal red and the underlying green or yellow shows clearly.

Color pattern *green-yellow spotted-solid red:* This pattern has the solid red as in the preceding type, but the conditions of yellow and green are as in type *green-yellow spotted-red blotched.*

Color pattern *green-solid red* (fig. 9): This pattern has the entire leaf above and below of a uniform *violet carmine.* It differs from type *green-yellow-solid red* in the absence of any underlying yellow, and from type *green-red blotched* in having the epidermis completely red instead of red in blotches. The pattern is dull and dark, with a somewhat metallic luster, in marked contrast to the various patterns with yellow.

Color pattern *green-yellow-solid red upper center* (figs. 10 and 10a): This is a brightly colored and attractive pattern with a rather complicated arrangement of colors. The subepidermal colors of green center and yellow border are similar to the arrangement in the types *green-yellow-red blotched* and *green-yellow-solid red.* The epidermal red is, however, almost entirely confined to the upper surface; over the central green it gives a *greenish violet carmine* cast; over the bordering yellow it forms a band of *nopal red.* At the extreme margin it is absent, giving a narrow but irregular band of pure yellow. On the under surface there are only occasional small blotches of red. About the border the red of the upper epidermis shows through the yellow, giving pale pinkish tints, as shown in figure 10a. This type was derived from pattern *green-yellow-solid red,* with its complete covering of epidermal red, by the loss of red on the under surface and about the extreme margin of the upper surface.

Color pattern *yellow-green-solid red* (fig. 11): This pattern was derived from type *green-yellow-solid red* by a reversal in the relative position of the underlying green and yellow, the change being the same that gave type *yellow-green-red blotched* from pattern *green-yellow-red blotched.*

Color pattern *green-yellow* (fig. 12): This is a bright pattern of green center and yellow border with *no expression of epidermal red.* Some few internal or vascular strands of red may be seen. The pattern differs from that of type *green-yellow-red blotched* in having no epidermal red.

Color pattern *green* (fig. 13): A pattern of pure *spinach green* with no yellow and no epidermal red, but with a few streaks of red in the vascular strands or in the mesophyl. This pattern differs from the parental pattern *green-yellow-red blotched* in the loss of both yellow and epidermal red. The *green-yellow spotted* pattern (type 13 a, not illustrated) differs only in having yellow spots.

Color pattern *yellow-green* (fig. 14): This type has a green border and a yellow center, with no epidermal red. It differs from the *green-yellow* pattern in the reversed position of the two color elements and from pattern *yellow-green-red blotched* in having no epidermal red. As in the case of the *green-yellow* type, the pattern is bright and attractive.

Color pattern *green-solid red upper center* (fig. 15): A type that differs from type *green-yellow-solid red upper center* in having no yellow and hence is apparently bicolored on the upper surface. The center of the leaf is *violet carmine* and the marginal zone is pure green.

Color pattern *yellow-solid red:* This type has almost *uniform nopal red* color on both surfaces. It differs from type *green-yellow-solid red* in not possessing a dark red center and from type *yellow-red blotched* in having the epidermis completely red. In both patterns the green underlying a solid red epidermis is almost entirely absent.

Laciniate leaf shape (fig. 7): In marked contrast to the type of entire leaf illustrated in the figures showing the various color patterns is the deeply and irregularly cut and lobed types of leaf shape, the appearance and behavior of which will be specially discussed later.

All of these color patterns arose as sudden spontaneous bud-variations, with the single exception of the type *green-solid red upper center*, which is a pattern into which plants with the *green-yellow-solid red upper center* pattern fluctuated. Throughout this paper, as above noted, the term "bud variation" is, in all cases not otherwise qualified, applied only to a marked change that appeared suddenly and completely for a part of a plant, and which was fully in evidence in the leaves involved when they first unrolled. Gradual fluctuations also gave in numerous cases types *green-yellow spotted-red blotched*, *green-yellow spotted*, and *green-red blotched*. That is, these types appeared both by sudden and by gradual variations.

CONSTANCY OF THE VARIOUS PATTERNS.

To test the constancy of the types, the original as well as those derived from it by bud variations, successive generations of plants were grown from pedigreed cuttings. This tested the vegetative constancy of the pattern itself and enables one to make comparisons when the same pattern was derived from different lines.

The series of plants considered under any type pattern are in large measure a selected stock. When cuttings were made for the perpetuation of the pattern in a new generation, they were made from the plants most typical and constant for the pattern concerned. When a bud variation appeared, if the conditions were favorable, the parts possessing it were allowed to develop until there were several branches from which cuttings could be taken simultaneously. In such cases the selection of branches for the new type was a simple matter, as it depended on the taking of branches sharply distinct from the main part of the plant, which in most cases were as different as is shown in figures 21 and 24. When further cuttings were made for a new generation to perpetuate the type they were made from plants most uniform and constant (determined from the records) for the pattern in question.

Usually but three cuttings were made from a plant, and these were taken from branches most uniform and clearly conforming to the type.

It has already been noted and it will be very evident in the following pages that some plants showed fluctuating variations giving irregular or mixed patterns, or the pattern gradually fluctuated between two types or changed from one type to another. Except in two cases no attempt was made to secure new types by such fluctuating variations. These cases (clone 14 of table 2 and clone 13 of table 3) will be especially discussed later.

In numerous cases cuttings were made to give two types of patterns in the same plant. The constancy of the patterns could in these cases be studied with the two parts growing from the same root system and submitted to the same environmental factors.

Plants with yellow-red blotched pattern (fig. 1).—Seven cuttings pure for this pattern were made in the autumn of 1913. Six died soon after they were placed in the rooting-bench. The other lived and was grown until the autumn of 1914. This plant was somewhat greener during the winter, but at all times was decidedly more yellow than any plant of any other type. It was, also, smaller and less vigorous in its growth. Eight plants were grown as chimeras with one branch of *yellow-red blotched* pattern and one branch of a pattern with *yellow-green-red blotched*. On all these the branches of the part with pattern *yellow-red blotched* remained quite constant throughout the year and were at all times in marked contrast to the pattern of the other part. Two chimeras grown only during the summer of 1914 were likewise quite constant. While it is very difficult to obtain plants with this pattern from cuttings, the type remains quite constant when grown in chimeral association with branches having green tissue. On account of the difficulties of propagation this type has not been rigorously tested. The few plants grown gave no marked variations either as bud variations or as fluctuations.

Plants with green-yellow-red blotched pattern (fig. 2).—This is the pattern originally possessed by the two parent plants, Nos. 1 and 3. A total of 337 plants were grown from cuttings of this type. The data given in table 2 are summarized in four main clones. Plants of clone 11 all descended from branch 1 on plant 1. Plants of clone 12 were descended from branch 2 on plant 1. Plants of clone 3 were derived from the branches of plant 3 that were uniform for this pattern. The original branches from which the first cuttings were obtained were uniform for the *green-yellow-red blotched* pattern and all plants used as parent stock for later generations were selected as typical and most constant for the pattern. The entire six generations constituted a series of plants derived by continued selection.

The 45 plants of this pattern in clone 14 are especially interesting, as they constitute a test for this pattern when derived by a gradual fluctuation. As already noted, branch 4 of plant 1 possessed a decidedly green pattern with yellow blotches designated as *green-yellow spotted-red blotched*.   In the third generation of plants grown from this branch 8 plants gradually fluctuated during winter until they were uniformly *green-yellow-red blotched*.   Such an increase of yellow during winter was unusual and cuttings were made to test the constancy of the type thus derived.   The data for the 45 plants grown in three generations show that two-thirds of the plants were quite constant for the derived type. This proportion compares very favorably with that of the clones 11, 12, and 3, which were from the start selected from plants most typical for the type.

TABLE 2.—*Summary of plants with green-yellow-red blotched pattern.*

	Clone 11.	Clone 12.	Clone 14.	Clone 3.	Total.
Number of plants.	151	103	45	38	337
Number constant.	102	76	21	19	218
Number fluctuating in green and yellow . .	29	22	15	11	77
Fluctuations to laciniate leaf-shape.	. . .	1	. .	1	2
Plants giving bud variations.	22	12	10	9	53
Bud variations:					
Yellow-red blotched . .	3	1	3	. .	7
Green-yellow spotted-red blotched	4	2	[1]4	2	8
Green-red blotched	13	7	1	3	24
Yellow-green-red blotched .	3	. . .	2	1	6
Green-yellow-solid red	. . .	1	1	2	4
Green-yellow.	3	1	1	3	8
Green-yellow spotted	. . .	. . .	[2]2	. .	2

[1]All were cases of fluctuation confined to about half of a plant.
[2]Loss of epidermal red on two plants fluctuating from *yellow* to *yellow spotted*.

Of the total number of plants with pattern *green-yellow-red blotched* there were 218 that were at all times fairly constant and true to the type.   They were all somewhat fluctuating in respect to the relative amounts of green and yellow, but were all constant in possessing at all times a yellow border.

In figure 2 there is shown a leaf with the average development of the yellow border, although in this leaf the pattern is somewhat irregular. Figures 10, 12, 19, 20, and 26 show leaves classed as yellow-bordered; figure 20, however, shows fluctuation toward type *yellow-red blotched*, and 26 shows fluctuation toward a *green-yellow spotted-red blotched* pattern. Figures 17 and 27 are from leaves classed as having irregular patterns.

The 77 cases classed as fluctuations include: (1) 56 cases of decided increase of green during winter, followed by increase of yellow in summer, giving in most cases return to the type of the cutting; two of these also gave fluctuations to laciniate-leaf shape; (2) 9 cases of increase of yellow during summer (grown only during a summer from cuttings taken from plants that were greener during the preceding winter); (3) 7 cases of fluctuations that were not uniform on a plant, but gave leaves of the same age with different patterns so mixed that no sectorial distribution could be traced; and (4) 5 plants with green and yellow distributed irregularly.

Aside from the fluctuations in relative amounts of green and yellow, there was also much fluctuation in number and size of the blotches of epidermal red.   On some of the plants there was rather gradual increase or decrease both in number and size of these blotches, giving such differences in respect to red as are shown in figures 5, 17, 26, and 28. Such plants were, however, still considered as blotched in the summaries.   Selection of typical *red blotched* epidermis for various types has been directed to plants having the epidermal blotching as in figures 2, 5, and 6, rather than as in figures 23 and 28.

53 plants produced bud variations giving loss of green 7 times, loss of the yellow bordering-band 8 times, complete loss of yellow 24 times, reversal of the relative positions of the green and yellow 6 times (4 cases appeared in half of a leaf only, as in fig. 25), increase of red to complete epidermal red 4 times, complete loss of epidermal red 10 times. In clone 14, 4 plants gradually developed a *green-yellow spotted-red blotched* pattern in part of the branches.   On the basis of my descriptions the change was a fluctuating variation affecting only a few branches of a plant.   These 4 cases are not included in the summaries of bud variations.

In their extent the bud variations gave extremes in development of yellow, of green, and of epidermal red.   There were cases of nearly pure yellow and of absolutely pure green; there were cases of solid red epidermis and others with no red epidermis.

Furthermore, the changes in green and yellow or in epidermal red occurred entirely independently of each other.   In general, the different types of bud variations were quite uniformly distributed in the various clones.

The type *green-yellow spotted* was produced on 2 plants by a fluctuational increase of green after the loss of epidermal red had occurred.

In table 2, as in other tables, when the totals given for constant, fluctuating, and sporting plants exceed the number of plants grown, it shows that a certain number of the fluctuating plants produced also sharp, clear-cut bud variations.   Also, when the total of cases of bud variations exceeds the plants giving them, certain plants produced more than one bud variation.

For the purpose of establishing an index of the frequency of bud variation we may take the ratio of bud variations to the estimated number of buds developed.   Each plant produced an average of at least 200 branches which made sufficient growth to reveal the pattern of the leaves.   On this basis the index of total bud variation for this group was about 1 to 1,110.   The ratio of constant plants to fluctuating plants was almost exactly 3 to 1, not counting the plants with bud variations many of which were otherwise constant.

Plants with color pattern green-yellow spotted-red blotched and with uniformly entire leaves.—The plants grouped in this class (figs. 4 and 23) present perhaps greater diversity than those of any other type, embracing (1) plants with considerable yellow in scattered areas in all leaves, (2) plants with only slight amounts of yellow in scattered areas in nearly all leaves, and (3) plants with only a few leaves possessing yellow spots. Between the extremes there was every degree of variation and often all degrees would be seen at one time among the leaves of a single plant.

It is difficult in such plants to determine what constitutes a variation either as a fluctuating or a bud variation when it involves green and yellow. The cases given in table 3 are those in which an entire branch or a sector of a branch showed leaves that were uniform for a new pattern. Plants having irregular mixtures of leaves of equally different patterns were common. Such cases are of special interest, as are the

TABLE 3.—*Plants with entire leaves and pattern green-yellow spotted-red blotched (fig. 4).*

	Clone 11.	Clone 12.	Clone 13.[1]	Clone 14.	Clone 3.	Total.
Total number of plants	2	16	79	89	22	198
Number constant..	..	6	61	50	9	126
Fluctuations for green and yellow .	..	2	16	21	8	47
To type green-yellow-red blotched...	..	..	..	8	..	8
To mixed patterns................	2	2	1	4	4	13
To laciniate leaf.....	..	4	..	3	..	7
Total number plants giving bud variations	..	2	2	3	1	8
Bud variations:						
To yellow-red blotched	..	..	..	1	..	1
To green-yellow-red blotched ..	..	1	1	1	..	3
To green-red blotched...	..	..	..	2	..	2
To green-yellow spotted-solid red...	..	1	..	..	1	2
To green-yellow spotted............	..	..	1	1	..	2

[1] Pattern derived by fluctuating variation.

fluctuating variations in a seed progeny. They possess many similarities to cases of size inheritance described by Goodspeed (1912) and raise the question as to whether color heredity is not also quantitatively rather than qualitatively inherited.

As has already been noted, the parent plant (No. 1) had one branch (No. 14) with leaves *green-yellow spotted-red blotched.* All the 89 plants of clone 14 descended from this branch through 6 generations of selection. The plants of this pattern here given with clones 11, 12, and 3 were obtained from cases of bud variation from the type *green-yellow-red blotched* (see table 2). The 79 plants of clone 13, however, were derived from 5 plants that gave a fluctuating change from *green-red blotched* to *green-yellow spotted-red blotched.* This was a frequent fluctuation from the *green* plants especially of clone 13, as shown in table 5, and the yellow-spotted condition thus obtained was tested in four generations, comprising a total of 79 plants.

As a whole, there was a rather large proportion of the plants that remained within the type as classified, although there was hardly a plant grown during the winter that did not become somewhat greener during that period. 68 plants fluctuated in a marked degree; 47 of them were almost entirely green during the winter, but were again quite uniformly *yellow spotted* in summer, although some of these remained much greener in summer. None of the latter, however, could be considered as of the pattern *green-red blotched*.

13 plants fluctuated irregularly, giving mixed patterns, mostly of *green-yellow-red blotched*, *green-yellow spotted-red blotched*, and *green-red blotched*, all more or less intermingled among the various branches and on the same branch. These were not used as parent plants, but doubtless by selection it would be possible to obtain a marked degree of constancy for the irregular and mixed patterns, with, also, production of plants that would be uniform for various types.

The most uniform and marked of the fluctuations was the case of 8 plants of clone 14 which gradually became more yellow during the winter of 1912–13, until they were quite typical *green-yellow-red blotched*; 5 of these were used as parents of the plants of clone 14 already reported with table 3. The change in pattern arose as a gradual increase of yellow from various degrees of a *yellow spotted* condition to a well-defined *yellow bordered* pattern that was quite uniform for the entire plant, and which when tested in progeny was subsequently quite as constant as cases which arose by sudden variations. An analysis of the pedigrees of these 8 cases shows that all of these descended from only 3 of the 7 plants grown from cuttings of the original branch 14. This phenomenon of the appearance of the same variation in different plants that were derived from the same more remote ancestor is common and constitutes what we may call duplicate-reversions or variations.

Besides the fluctuating variations in regard to green and yellow, there were numerous cases of fluctuation in the red-blotched condition both of the epidermis and of the subepidermal tissues, giving extremes of very finely red-blotched or coarsely blotched. No selections were made to secure types of the red-blotched condition. None of the plants fluctuated to a no-blotched or to a solid-red pattern.

The behavior of the 79 plants of this pattern in clone 13 is especially interesting. They constitute a test for this pattern obtained by the selection of gradual and accumulated fluctuation. The progeny of 5 plants grown in 3 generations, subjected to the same sort of selection as the other clones, showed the highest degree of constancy obtained in the clones of this pattern.

Seven plants gave, during the winter, a marked increase of green, accompanied by the production of cut and laciniate leaves (fig. 7), the appearance and constancy of which are quite fully discussed later.

There were few cases of sudden, clear-cut bud variation in this group. One was a very decided and almost complete development of yellow to *yellow-red blotched;* 3 were sectorial variations to *green-yellow-red blotched;* 2 cases gave loss of yellow to *green-red blotched;* 2 involved gain of epidermal red to solid red, and in two cases there was loss of epidermal red. All these cases were sectorial for a plant or in some cases for a single branch. The ratio giving the frequency of bud variations involving color in this group is 1 to 3,960.

Plants with laciniate leaves.—Until the winter of 1912–13, all the plants in the cultures had been constant and very uniform for leaf-shape, showing no greater variation in this respect that is seen in figures 2, 6, and 9. During that winter it was noted in 11 cases that as new leaves developed they were more and more deeply cut and lobed until in January and February the uppermost leaves were in extreme cases much divided and deeply laciniate, as shown in figure 7. The plants appeared like the middle plant in plate 4; 3 of these plants were from cuttings of branches that were pure-green bud variations; the others were plants which had a somewhat fluctuating *green-yellow spotted* pattern. Nine of these plants were grown during the following summer, when it was noted that without exception the new leaves gradually became more entire until by late summer all the leaves then hanging to the plants were entire. At the same time 5 of the plants became decidedly more yellow, even becoming quite uniform for type *green-yellow-red blotched.* Cuttings were made in April from each of these plants and from these 16 plants were grown during the summer of 1913. All of these fluctuated to entire leaves, and in regard to color gave plants some of which were uniform for type *green-yellow-red blotched,* others for type *green-yellow spotted-red blotched,* while the foliage of others showed mixtures of these patterns with also pattern *green-red blotched.*

The further generations in subclone 12 exhibited, as shown in table 4, the same periodicity in change of leaf-form, except that in late summer of 1914 a rather large number of the plants showed new leaves that were laciniate. One plant of this subclone was grown in the winter of 1913–14 from a pure-green bud-sport that appeared during the previous summer. This plant remained constant for entire leaves of pattern *green-red blotched* during the winter, but as it died early in the summer no further progeny were grown. Further generations of the plants with laciniate leaves in subclone 14 were not grown.

The laciniate leaf-form appeared anew in the winter of 1913–14 in clone 3 in two instances. One was a plant whose line of descent showed bud variation from type *green-yellow-red blotched* to type *green-yellow spotted-red blotched;* that of the other showed bud variation from *green-yellow-red blotched* to *green-yellow solid red,* and then from this to *green-yellow-solid red upper center.* The latter was the only one of

8 plants grown from cuttings of a single plant to exhibit the fluctuation to laciniate leaves.

It is to be noted that this character of laciniate leaf-shape has not appeared thus far in any of the plants grown in clones 11 and 13. It has appeared as a fluctuating character that develops most strongly in winter. With one exception all the plants grown to test the reappearance of the character have exhibited it. This plant was grown from a bud variation giving a single branch of *green-red blotched* on a plant otherwise uniform for pattern *green-yellow-red blotched* at the time the cutting was made.

During the time these plants exhibited the laciniate character most strongly, there were growing among them numerous plants of other clones of various patterns, especially of *green-yellow-red blotched* and *green-red blotched*, all submitted to the same conditions of light, tem-

TABLE 4.—*Summary of plants with laciniate leaves.*

	Clone 12.				Clone 14.		Clone 3.	
	1912–1913.	1913.	1913–1914.	1914.	1912–1913.	1913.	1913–1914.	1914.
Total number of plants. . .	6	10	32	7	3	6	2	2
Much greener in winter . . .	6	..	32	..	3	..	2	..
Very laciniate in winter . .	6	..	31	..	3	..	2	..
Entire in winter..............	0	..	1	..	0	..	0	..
In summer as type 2.	5	4	4	1	0	0	0	0
as type 4..............	0	1	5	3	0	6	0	1
as type 1..............	0	0	1	0	0	0	0	1
mixed patterns . . .	1	5	22	3	3	0	2	0
uniformly entire . .	6	10	5	0	3	6	1	0
slightly laciniate .	0	0	27	7	0	0	1	2
Plants giving bud variations.	1	..	1	..	..	1	..	..
Type green-red blotched and entire . .	1	..	..	..	..	..	..	..
green-yellow spotted	..	..	1	..	..	..	..	..
green-yellow-red blotched	..	..	..	..	..	1	..	..

perature and soil, yet not in the least degree exhibiting the fluctuation to laciniate leaves. This is well shown in plate 4, which gives a photograph of 3 plants of the same clone (12), all grown under the same conditions. The plant to the right (No. 125111) had entire leaves and a *green-yellow-red blotched* pattern; the one to the left (No. 1251412) was of a pure *green-red blotched* pattern; the one in the middle (No. 123153) shows the transition from entire leaves to deeply laciniate leaves as it occurs during the winter.

The late summer of 1914 was exceedingly dry. In July there had been 5.36 inches of rainfall, during which time the plants made an unusually vigorous growth. From August 12 until October 16 there was but 1.26 inches of rain. During the dry warm period of September, the new leaves on very many of the plants of this series were strongly laciniate. Cuttings were made from these laciniate-leaved branches. The new leaves that developed on these young plants during November were entire; hence it would seem that the laciniate character in these particular clones of *Coleus* is in some degree associated with decreased vigor. When most favorable conditions for growth prevail, or when rapid growth is brought about in cuttings, the leaves become entire.

It is, however, clear that the first appearance of the laciniate character was confined to a few plants and that once it originated it reappeared with marked constancy in the vegetative progeny.

During the winter of 1914–15 the laciniate character appeared in the manner of a bud variation. A large plant that had been grown out of doors during the summer was in September severely pruned and placed in a pot for further development in the greenhouse. It was intended to use the plant for stock in general border planting and the plant label was not preserved. From the records of the pattern and generation it is clear that the plant itself and all the plants in its line of descent possessed only entire leaves and the plant belonged to the main clone 1. This plant was given the number 9.

In the course of 3 months numerous new branches developed on the 10 pairs of main lateral branches to which the plant had been pruned. It was noticed that of the 20 main branches, 3 bore branches with laciniate leaves. The positions of these are indicated by numbers 1, 2, and 3 of diagram 2.

All the branches on all other of the main branches bore only entire leaves. All the branches arising from 2 bore only laciniate leaves, but the branches with entire and with laciniate leaves were sectorially distributed on the branches 1 and 3. The contrast was most marked, especially when opposite branches were different, one having laciniate and the other entire leaves.

10		10′
	9	9′
8		8′
	7	7′
6		6′
	5	5′
4		4′
	3	3′
2		2′
	1	1′

DIAGRAM 2.—Showing position of the branches on plant No. 9.

The sectorial differences appeared in some of the secondary branches and carried the two types into parts of individual leaves.

The most striking behavior of this series of plants summarized in table 4 is the wide fluctuations in the leaf-shape and in the amount of yellow and green, the marked correlation of decrease of yellow with decrease of leaf area, and the rather pronounced periodicity of these fluctuations. These fluctuations are so general and rhythmic that they can almost be considered constant. Of bud variations there were but 3 cases, giving a ratio of about 1 to 2,530, which, however, shows greater frequency than that of the pattern *green-yellow spotted-red blotched*. Of the bud variations one was a loss of yellow, one gave increase of yellow to type *green-yellow-red blotched* and one was a loss of red.

Plants with green-red blotched pattern.—The apparently complete loss of yellow, giving only green subepidermal tissues, was a frequent bud variation in plants having *green-yellow, yellow-green,* or *green-yellow spotted* patterns, regardless of the degree of red in the epidermis (fig. 5). The condition of pure green also developed as a fluctuation on plants of these same types. In cases the fluctuation was quite general for the entire plant, while in others it occurred irregularly, giving plants with mixed patterns.

The 90 plants included in this summary are, however, selected stock, all descended from cases of bud variation similar to that of figure 21, in which the part concerned showed no trace of yellow in any leaves.

TABLE 5.—*Summary of plants with green-red blotched pattern (fig. 5).*

	Clone 11.	Clone 12.	Clone 13.	Clone 14.	Clone 3.	Total.
Number of plants	13	18	51	6	2	90
Plants constant.	5	5	17	3	1	31
Fluctuations:						
Green-yellow spotted	1	7	31	1	1	41
Mixed patterns	..	1	3	..	..	4
Laciniate leaf-shape	..	3	..	..	..	3
Plants with bud variations.	7	5	1	2	..	15
Bud variations:						
Green-yellow spotted-red blotched	1	1	..	..	..	2
Spontaneous yellow	6	1	..	2	..	9
Green-solid red	..	2	..	..	..	2
Green	1	1	1	..	..	3

Selections for further generations were made from plants that had remained uniformly pure green. The type was maintained by selection quite as it is practiced in a herd of dairy cattle. In the case of clone 13, 6 generations were grown, all descended from the pure-green branch of the parent plant, No. 1.

Of the total number of plants in this group, 31 remained pure green, showing no trace of yellow by either fluctuation or bud variation. In addition, 14 of the 15 plants with bud variations were otherwise constant for the pure-green condition. All of these plants were grown during an entire summer. 41 plants developed varying amounts of yellow in scattered areas, making a pattern classed as *green-yellow spotted-red blotched*. In 11 of these the *yellow spotted* condition was quite uniform and typical and from these were selected parents for the plants of subclone 13 given in the table 3. Three plants also gave fluctuations in leaf-shape to the laciniate type and their progeny are included in the summary of table 4. Four plants gave decidedly mixed and irregular patterns like those that appeared from patterns treated in tables 3 and 4.

There were 16 cases of bud variation; 5 were concerned with the red epidermis, 2 giving solid red, and 3 giving no red. There were 2 cases of spontaneous appearance of yellow, giving branches sectorial for the *green-yellow spotted-red blotched* pattern. There were, also, 9 cases of spontaneous development of yellow that were not carried on in successive leaves, and with the exception of 1 case were confined to but one or two leaves. These yellow blotches were large, irregular-shaped pure-yellow areas covering from one-eighth to one-fourth the entire area of a leaf. The locations and relative sizes of these yellow blotches in the leaves of one plant are shown in text-figure 1, the shaded protions of which indicate yellow areas. All other leaves were pure green and the branches produced in the axils of the yellow-blotched leaves were pure green.

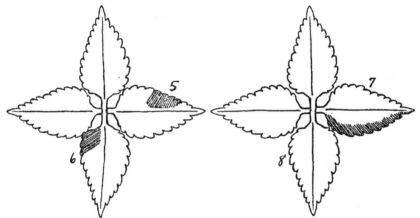

TEXT-FIGURE 1.—Position of the yellow areas that developed spontaneously in these leaves of plant 11714221.

A study of pedigrees reveals the interesting fact that 6 of the plants with spontaneous yellow all descended from a branch on plant 11714. The full record of this clone is given in table 13, but the summary of the pedigrees of the particular line of descent is here given in table 6.

The pedigree numbers enable one to trace relationship quite readily. In this case we note that plant 1171 gave a bud variation with loss of yellow. Plant 11714 was grown as a chimera with about one-half pure green. From the green part two cuttings were made for plants 117142 and 117144, both of which remained constant for loss of yellow. In September 1913, three cuttings were made and from two of the plants grown three more cuttings were taken in the following spring. From September 1912 until August 1914 all plants grown in these lines of descent were constant for loss of yellow, then in one season the spontaneous development of yellow occured in 6 closely related plants.

TABLE 6.—*Pedigree of 6 plants giving spontaneous yellow.*

Time.	Plant.	Pattern.	Record.
Apr. 1912 to Sept. 1914	117......	Green - yellow - red blotched...........	Constant.
Sept. 1912 to Sept. 1913	117, 1.....	... do..............	Bud variation to green-red blotched.
Summer 1913.........	117, 14[1]...	{ .. do.	Constant.
		{ Green-red blotched...	Constant for loss of yellow.
	117, 142....	.....do..............	Do.
	117, 144....	.....do..............	Do.
Sept. 1913 to Sept. 1914	117,142,1...	. do.........	Spontaneous yellow in August.
	117,142,2...	.. ..do....	Do.
	117,144,1...	.....do	Do.
Summer 1914.........	117,142,11..	.....do..	Do.
	117,142,12..	..... do.	Do.
	117,142,21..	.....do.............	Do.

[1]Chimera.

Including the cases of spontaneous appearance of yellow with sectorial bud variations, the ratio of frequency was 1 to 1,120. The ratio for the sectorial bud variations alone was about 1 to 2,570.

Plants with pattern yellow-green-red blotched.—All of the plants grown with this pattern (fig. 6) belonged to clone 11, and all, excepting 3, descended from the plant 1171, which produced a series of branches with this pattern, as already shown in diagram 1. The record of these plants is given later in table 13, but may be summarized here as given in table 7.

TABLE 7.—*Summary of plants with yellow-green-red blotched pattern (fig. 6), all of clone 11.*

	Total number of plants.	Constant plants.	Fluctuating.	Plants giving bud variations.	Bud variations.			
					To yellow-red blotched.	To green-yellow spotted-red blotched.	Green-red blotched.	Yellow-green.
Yellow-green-red blotched....	29	17	3	9	3	1	6	1
Chimera { yellow-green-red blotched.. / yellow-red blotched........	8	5	1	2	..	..	2	..
Chimera { yellow-green-red blotched... / green-red blotched.........	2	2	..	..	..	..	..	..
Chimera { yellow-green-red blotched.. / yellow-green.............	2	..	..	2	..	..	2	..
Total.......................	41	24	4	13	3	1	10	1

In several instances cuttings were so made that the resulting plants possessed two patterns. These plants are included here in respect to the behavior of the part with pattern *yellow-green-red blotched*.

All of the plants grown during the winter showed more or less increase of green, but as long as the yellow was present as a definite central area they were classed as constant. There were, for example, just such differences in development of yellow as is seen in figures 14 and 14*a*, the former representing the usual condition during winter and the latter the development of yellow during the summer. Four plants fluctuated in marked degree, giving mixed and irregular patterns with many leaves in which there was much green.

One case of bud variation was concerned with loss of epidermal red, giving the type *yellow-green* (fig. 14). The other instances gave 3 cases of extreme development of yellow, 10 cases of entire loss of yellow, and 1 case of change to the *yellow spotted* condition. The latter, however, occurred on a plant with also a bud variation to pure *green-red blotched*. The ratio of frequency for bud variation in this group was about 1 to 540.

Plants with pattern green-yellow-solid red.—In respect to the green and yellow this pattern (fig. 8) is identical with that of *green-yellow-red blotched*. It differs in having a solid-red instead of a red-blotched epidermis. The 54 plants grown with this pattern remained free from any noticeable variations in respect to the *solid red* epidermis, except those that were bud variations. Frequently a leaf appeared with a few small areas in which the red of the epidermis was absent, but these were rather isolated. There was some degree of fluctuation in the relative amounts of green and yellow, with a tendency for plants to be greener in winter and yellower in summer. On account of the solid red, it was more difficult to judge these fluctuations than in plants with *red blotched* or with *non-red* epidermis, hence attention was chiefly directed to the condition of the epidermis. In making cuttings, plants most constant and typical for the *green-yellow* condition were, however, selected.

Of the 10 cases of bud variation, 4 gave complete loss of yellow, 2 gave a reversal of the relative position of green and yellow, and 1 gave extreme development of yellow. Only 3 cases involved variation in the amount and distribution of red; 1 was a complete loss and 2 gave the pattern described above as *green-yellow-solid red upper center*, a type which is an interesting intermediate between *no red* and *solid red*. The ratio of frequency for all bud variations for the group is 1 to 1,080.

Plants with green-solid red pattern.—This pattern (fig. 9) first appeared during the summer of 1913 as a bud variation on a plant of pattern *green-red blotched*. From this branch cuttings were taken for 6 plants grown during the summer of 1914. All of these remained constant for loss of yellow and for a solid-red epidermis, except 1 plant, upon which a sectorial variation in one small branch gave the pattern *green*.

With respect to the red epidermis, patterns *green-yellow-solid red* (fig. 8) and *green-solid red* are identical, and it is noteworthy that in the 62 plants of these two patterns which were grown to maturity there were no noticeable fluctuations and but 4 cases of bud variations involving the red. For the red epidermis the ratio of frequency of bud variations was 1 to 3,100.

TABLE 8.—*Summary of plants with solid red patterns, clone 3.*

	Green-yellow-solid red.	Green-solid red.
Total number of plants...............	54	8
Constant for solid red...............	51	7
Number of plants with bud variations...	9	1
Bud variations:		
To yellow-solid red.................	1	..
To yellow-green-solid red...........	2	..
To green-yellow-solid red upper center.	2	..
To green-solid red....:.............	4	..
To green-yellow....................	1	..
To green.........................	..	1

Plants with pattern green-yellow-solid red upper center.—This bright and attractive pattern (fig. 10) first developed as a sectorial bud variation in the winter of 1912 on a plant otherwise uniform and constant to type *green-yellow-solid red*. During the summer of 1913, the plant grew vigorously and numerous branches developed from the part having this new pattern, all of which were constant and uniform for the new type and which were in conspicuous contrast to the rest of the plant.

In the autumn of 1913 cuttings were made from these branches for 5 plants, which were grown until the autumn of 1914. Four of these remained quite constant, although they were much greener in winter. One plant became gradually greener during the early part of winter until it was apparently pure green, giving type *green-solid red upper center*. The upper surface of a leaf of this plant painted in January is shown in figure 15. Toward spring the new leaves produced by this plant became quite laciniate, but during the following summer the leaves produced were entire and strongly tinged with yellow. During the summer of 1914, two plants from cuttings of one of the plants constant for the type remained true to that pattern.

No bud variations appeared in any of the 7 plants and no noticeable fluctuations in the amount and distribution of the red; yet there was no plant that did not show at some time a few leaves with tiny red spots scattered on the lower surface, much as is shown in figure 10a.

This pattern also appeared late in the summer of 1914 as a sectorial variation on a plant which during the summer had been constant and uniform for type *green-yellow-solid red*. A cutting was made from this branch, and the young plants grown from it are at the present writing (December 10, 1914) nearly devoid of yellow, but have the change in pattern for red as a clear-cut sectorial variation. Both surfaces of a single leaf are shown in figures 24 and 24a. In figure 24 the loss of epidermal red on the lower surface of half of the leaf illustrates very well the definiteness with which color variations in *Coleus* appear. The upper surface of the corresponding half of this leaf is shown in figure 24a, with the decrease of red about the margin. Such differences are usually seen in a series of leaves in the same row and in the branches that develop in the axils of such leaves, giving a marked degree of sectorial symmetry to the distribution of pigmentation, a condition also well illustrated with reference to green and yellow in figure 21.

Plants with pattern yellow-green-solid red.—This pattern (fig. 11) first appeared during the summer of 1913 as a variation complete for a single lateral branch of a plant with *green-yellow-solid red*. This branch was removed for a cutting, but died soon after it was rooted and placed in a pot. Early in the spring of 1914 this pattern appeared as a sectorial variation in the main axis of a plant having *green-yellow-solid red*. This plant grew vigorously, giving a large, bushy plant with the two types of foliage distinct and constant on the different branches. Numerous cuttings have been made to test the vegetative constancy of this type.

Plants with pattern green-yellow.—Three plants of this type (fig. 12) were grown from September 1913 until October 1914 and 4 were grown during the summer of 1914. All of these remained quite constant for the loss of epidermal red. They were much less uniform in regard to the relative amounts of green and yellow, one plant possessing a branch that was quite green. There was also a strong tendency among the leaves on one plant to show somewhat irregular distribution by green and yellow, as is shown in figure 12.

Plants with pattern green and pattern green-yellow spotted.—The loss of epidermal red occurred as a sectorial bud variation during the late summer of 1913 on a plant that had fluctuated from type *green-yellow-red blotched* to type *green-yellow spotted-red blotched*. The development of yellow was, however, very faint, so that the bud variation gave a leaf pattern that was almost pure green. This plant was taken up and grown in a pot during the winter of 1913–14 and during the following summer again grown out of doors. All branches on the two parts were quite constant in respect to presence and absence of the epidermal red, but there were traces of red coloration in the sub-epidermal tissues which, as shown in figure 13, were almost entirely confined to the vascular tissues. There was more or less fluctuation on the entire plant in the appearance of yellow, but no decided development of it.

In January 1913 a cutting was made from the part of this plant which showed the bud variation. The plant grew vigorously and was during the winter mostly free from yellow. In April, 5 cuttings were made for plants. During the summer of 1914 the 6 plants remained constant for loss of epidermal red, but bore some leaves with yellow spots. There was, however, a rather weak development of yellow and from a short distance the plants appeared to be pure green. The fluctuations in the development of green and yellow are quite like those in other patterns. The pattern is of special interest in regard to the development of red pigmentation in the subepidermal tissues, especially in the vascular elements giving a reticulated effect well shown in figure 13, and also seen in figure 14a. This condition also prevails in the red blotched types, but is more or less obscured by the more conspicuous epidermal coloration.

During the summer of 1914 there was no noticeable variation in the amount of red in subepidermal tissues. Cuttings made in the autumn of that year for a new generation exhibited during the winter marked variations in this respect. The summaries given in this paper do not include the generation to which these plants belong, but the behavior of this particular set of plants can be included here. Figure

TABLE 9.—*Summary of plants with non red epidermis.*

Pattern.	Total plants.	Constant for non red epidermis.	Plants giving bud variations.
Green-yellow..............	7	7	0
Green and green-yellow spotted....	7	7	0
Yellow-green....	4	4	0

13d gives a leaf painted on February 2, 1915, showing the development of red in the internal tissues. Free-hand sections of such leaves indicated that the epidermal cells are non-red. The coloration appears dull, as if glazed over rather than velvety as in the epidermal coloration, a contrast due largely to the coloration of the trichomes of the epidermis and which the reproductions do not adequately show.

Plants with pattern yellow-green.—This pattern (fig. 14) identical with that of type *yellow-green-red blotched* except for the loss of epidermal red, appeared as a sectorial bud variation late in the summer of 1913. The sporting branch was used as a cutting, from which a large plant grew during the winter of 1913 and the following summer. In the winter there was an increase of green, but throughout its growth the part with pattern *yellow-green* remained constant in respect to the loss of epidermal red.

Three plants of a new generation were grown during the summer of 1914, and these remained constant and uniform for the loss of epidermal red. In regard to the relative positions of green and yellow, the plants were quite constant, but there was a strong tendency for green to increase in winter and decrease in summer, giving such differences as are shown in figures 14 and 14a.

The figures 13, *a, b,* and *c,* show the three leaves growing in the rank above the leaf shown in figure 13*d* and illustrate the increase of red which very plainly occurs as the leaves mature. In the *red-blotched* and *solid red* patterns an increase in the total amount of red pigmentation must, it would seem, also occur as the leaves enlarge.

As the last four patterns, *green-yellow, green, green-yellow spotted,* and *yellow-green* are alike in respect to loss of epidermal red, they may be grouped in this respect. It is noteworthy that there was no case of a development of epidermal red. No plant was free of some red coloration in stems and in vascular strands of the leaves, as especially well shown in figures 13 and 14. There were cases where the coloration seemed to spread out near the ends of the vascular strands, but the appearance was not the same as that of the blotches. The number of plants of these patterns grown thus far is small and their behavior is not taken as fully indicative of the possible variations that may appear in future cultures.

From the summaries of patterns given above it is quite clear that the various types noted (with the exception of pattern *green-solid red upper center*) have been kept quite constant by a selection of the parent plants to be used in vegetative propagation, and that every new type of pattern (excepting the one) that arose either by fluctuating variation or by bud variation can be propagated as a vegetative type. It is highly possible that finer distinctions could be made in regard to pattern types, especially inside of the rather comprehensive groups classed as *red blotched,* as *yellow bordered,* and as *yellow spotted,* among which there were many variations that gave all degrees of gradation to or even into a different pattern. The writer wishes to state that the keeping of records satisfactory to him was no simple matter, even for the pattern classes as determined.

RANGE OF THE VARIATIONS.

At this point the data already presented may be summarized and analyzed in reference to the range or extent of the variations as a whole.

There have already been described 15 different color patterns that arose by bud variation, 1 very decided color pattern that arose solely as a fluctuating variation, and the laciniate type of leaf. Several of the principal types arose also by fluctuating variations. In respect to the relative amounts of green and yellow there are the two extremes: (a) almost pure development of yellow (fig. 1), and (b) pure green (figs. 5 and 9), with almost every possible gradation between. Of the epidermal red there are the extremes *solid red* and *no red* with the intermediate *red-blotched* type (including wide variations), and the type *solid red upper center* as another intermediate. In respect to the relative positions of the green and yellow there are the extremes: (a) green center with yellow border (fig. 2) and (b) yellow center with green border (fig. 6). Between these the irregular patterns present numerous intermediates.

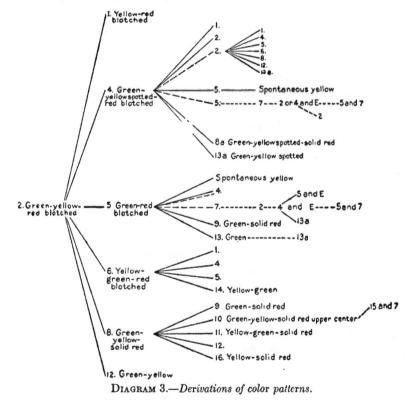

DIAGRAM 3.—*Derivations of color patterns.*

Diagram 3 gives a graphic representation of the extent of variations in each pattern and shows the derivation of types and the appearance of the same types of pattern as variations from quite different patterns.

In diagram 3 the numbers (except 13 a) refer to figures illustrating the types of color pattern or leaf-shape. The name of the pattern is written in full only in the line of descent when it first appeared. A continuous line indicates origin by bud variation, while a broken line indicates fluctuation.

From the original pattern of *green-yellow-red blotched* there arose directly 6 different patterns (see tables 1 and 2 and diagram 3) each involving a single marked variation. For the development of any other pattern thus far produced, excepting the *green-yellow-solid red upper center* pattern a second change is necessary. On any of these 6 derived patterns a further single change may give a new pattern or produce a pattern already realized. The bud variations in the derived pattern *green-yellow spotted-red blotched* (fig. 4) illustrate this point. Increase of yellow gave *yellow-red blotched* (fig. 1) and loss of yellow gave *green-red blotched* (fig. 5), both patterns previously derived, and also a return to the original type of *green-yellow-red blotched* (fig. 2). In all these the red-blotched condition of the epidermis is much the same. Changes in the epidermal coloration, however, give new patterns. The appearance of solid red gives a slightly different pattern than that of *green-yellow-solid red*. The pattern *green-yellow spotted* appeared as a bud variation by a loss of the epidermal red.

The *green-red blotched* (fig. 5) pattern gave opportunity for new patterns by the same changes in epidermal red which have previously appeared. These are realized in *green-solid red* (fig. 9) and *green-no red* (fig. 13). There is also chance for reappearance of yellow to give return to old types or possibly to new types. Of these only one appeared, and this was the pattern *green-yellow spotted-red blotched* (fig. 4). The cases of spontaneous appearance of yellow did not result in a definite pattern. In this line of descent there also developed the character of laciniate shape of the leaf, with its marked periodicity of expression.

The changes in the pattern *yellow-green-red blotched* (fig. 6) which involved amounts of green and yellow gave no new patterns. The loss of red, however, gave a new pattern *yellow-green* (fig. 14). At least 3 possible bud sports did not appear in this pattern: (1) changes producing a reversal of green and yellow giving return to the parent type; (2) a *solid-red* (fig. 11); or (3) *solid red upper center* (fig. 10).

In the plants with *green-yellow-solid red* (fig. 8), a loss of yellow gave the same pattern that was produced by gain of solid red from pattern *green-red blotched*. The two changes involved are identical, but occurred in reversed order. Reversal of the positions of green and yellow, a change identical with that giving the *yellow-green-red blotched* pattern, gave a different pattern because the tissues were overlaid by solid red. The same is true of the loss of green. Loss of the epidermal red on the lower surface and about the margin of the upper surface gave a pattern *green-yellow-solid red upper center* (fig. 10) that has not appeared elsewhere and is really the only new variation that appeared in this pattern.

Plants grown from pattern *green-yellow-red blotched* (fig. 2) derived from *green-yellow spotted-red blotched* (fig. 4) gave again the same patterns that were directly derived from that pattern.

A general review of the entire series of variations shows that in respect to the relative positions and total amounts of green and yellow, and the total amount of red in the epidermis, the extremes of development possible are realized, with, also, the appearance of a series of intermediate types. Judging the variations in any one pattern by the range of bud variations that have thus far developed, it appears that any pattern tested in considerable numbers gives by bud variation the entire range of changes possible.

FREQUENCY OF THE BUD VARIATIONS.

With the list of the types of variations given in table 1 there is also given the data as to the number of times each occurred and the total number of plants involved (not including plants of patterns which did not produce the particular variation). The various tables present the details of their data, which may be now summarized under the main types of changes outlined in table 1.

A. CHANGES INVOLVING YELLOW AND GREEN.

I. INCREASE OF YELLOW AND DECREASE OF GREEN.

1. The almost complete loss of green with increase of yellow occurred as a bud variation 12 times in a total of 630 plants (not including plants of patterns which did not give this variation). It was derived most frequently from patterns *green-yellow-red blotched* (fig. 2), *yellow-green-red blotched* (fig. 6), and *green-yellow-solid red* (fig. 8). It developed once from pattern *green-yellow spotted-red blotched* (fig. 4). The pattern was not realized uniformly on any plant as a fluctuating variation. On a few plants of the pattern *green-yellow-red blotched* which developed mixed patterns with a marked increase of yellow during the summer, some leaves approached this pattern. One of the most marked of these is shown in figure 20.

2. It will be remembered that the *green-yellow-red blotched* pattern was borne by the parent plants and that all other patterns were derived directly or indirectly from this. The return to this pattern occurred as a bud variation from type *green-yellow spotted-red blotched* in 4 instances on a total of 266 plants and also as fluctuating variations, especially in clone 14, as shown in table 3.

3. The sudden appearance of scattered areas of yellow in single branches of plants otherwise having no yellow occurred but twice. The same type *green-yellow spotted-red blotched*, however, appeared quite gradually for entire plants in 38 cases out of the 90 plants which were grown for the pattern *green-red blotched*.

4. The development of rather large conspicuous areas of yellow in one or more leaves of a plant otherwise pure green occurred 9 times. In one case three leaves of the same branch showed yellow areas that arose in this apparently spontaneous manner (see text-fig. 1).

Summary.—There were 27 cases of bud variation giving increase of yellow and involving directly 788 plants. There was opportunity for increase of yellow to occur in all plants grown, except those of the pattern *yellow-red blotched,* and even in these, 10 of the 11 were grown as chimeras with at least half of the plant green. In computing a final ratio for the frequency of bud variations giving increase of yellow, we may use all but 6 of the plants grown. The ratio of frequency on this basis is 1 to 6,130.

II. Decrease of Yellow and Increase of Green.

1. Pattern *green-yellow spotted-red blotched* was produced from *green-yellow-red blotched* and *yellow-green-red blotched* in 9 instances on a total of 378 plants. In 4 other cases the change to *yellow spotted* affected single branches, and although marked for a time after the first appearance, later fluctuated to the parent type *green-yellow-red spotted,* for which the plants became quite uniform.

2. Pattern *green-red blotched* with complete loss of yellow occurred on plants with *green-yellow-red blotched,* with *green-yellow spotted-red blotched* (entire and laciniate), and with *yellow-green-red blotched* patterns in 37 instances on a total of 644 plants. The same change gave pattern *green-solid red* 4 times on 54 plants of pattern *green-yellow-solid red.*

Summary.—Bud variations producing increase of green occurred 50 times. The total plants grown with more or less yellow were 740. The ratio of frequency for loss of green by bud variation was 1 to 2,960.

III. Reversal of the Relative Positions of Green and Yellow.

1. This reversal has only occurred in patterns with the yellow at the border of the leaf, giving *yellow-green-red blotched* (fig. 6) from *green-yellow-red blotched* (fig. 2) and *yellow-green-solid red* (fig. 11) from *green-yellow-solid red* (fig. 8) in a total of 8 instances on 391 plants. It is also possible for a reversal to occur in any other patterns having a distinct border of green or yellow. The total of such plants is 450, which gives 1 to 11,250 as the ratio of frequency for this change.

B. CHANGES INVOLVING THE EPIDERMAL RED.

I. Increase of Epidermal Red.

Eight instances of bud variations giving solid red occurred in red blotched patterns involving directly a total of 625 plants. None of the 41 plants of the pattern *yellow-green-red blotched* gave this variation. Increase of red was possible in all except the solid-red patterns (62 plants in all). The ratio for this change was 1 to 19,250.

II. Decrease of Epidermal Red.

Bud variations of this sort can be graded as follows: an almost complete loss of red on the under surface and about the upper margin, which occurred 2 times, and an apparently complete loss of epidermal red in patterns with spotted or with solid red epidermis, which appeared in 19 instances. A total of 815 plants were grown of patterns having some degree of red in the epidermis. The ratio for complete loss of red was 1 to 8,580 and for all cases of decrease of red it was 1 to 7,760.

C. CHANGES INVOLVING LEAF-SHAPE.

The appearance of the laciniate leaf-shape as a fluctuating variation which marked periodicity of development occurred 13 times. (The bud variation giving this type late in 1914 is not included.) The total number of plants grown with entire leaves was 765, hence the ratio on the basis used hitherto was 1 to 11,770. It seems, however, that this basis hardly affords the same degree of accuracy for comparison as it does between the different bud variations in color. Here the change appeared in an entire plant (except one plant grown during winter of 1914–15 and not included in these computations), but as several of these were from a same immediate parent, it may be that the change really arose as a bud variation, with, however, a delayed effect.

TABLE 10.—*Frequency of changes giving the different types.*

Type of change.	Plants.	Frequency.	Ratio.
Increase of yellow and decrease of green	827	27	1 : 6,130
Decrease of yellow and increase of green	740	50	1 : 2,960
Reversal of positions of green and yellow. . . .	450	8	1 : 11,250
Increase of epidermal red to solid red	770	8	1 : 19,250
Decrease of epidermal red, complete loss. . . .	815	19	1 : 8,580
Decrease of epidermal red, all cases	815	21	1 : 7,760
Appearance of the laciniate character	765	13	1 : 11,770
Entire leaf from laciniate leaf	68	1	1 : 13,600

Of the 68 plants grown with the habit of producing laciniate leaves, a single case of persistent change to the entire leaf-shape appeared. There were also 3 cases of clear-cut bud variations involving color changes in these plants.

SUMMARY AND COMPARISONS.

For the purpose of comparison, the ratios showing the frequency with which these different types of changes appear are brought together in table 10. In deriving these ratios the total number of plants in which there is possibility for the change to occur has been considered.

These data indicate the tendencies of the bud variations and give a clew to the behavior of the characters in question. In the bud variations, decrease of yellow occurred twice as often as the increase of yellow. Likewise, the loss of red occurred 2.2 times as often as the increase of red. If we consider that the ability to produce yellow and red are the more recently acquired characters of the cells, these data would indicate a tendency toward loss rather than gain of these characters.

A summary of the data regarding the degree of constancy of the various patterns and the nature of the variations which they exhibit is of further interest in a consideration of the tendencies of the variations. In comparing bud variations which originate .in a bud the comparison on the basis of the total buds produced seems quite adequate. The comparison of fluctuating variations requires a different treatment. On plants with irregular and mixed patterns it is not practicable, if possible, to attempt a statistical determination of the fluctuating branches. Only in few cases when such changes were

TABLE 11.—*Summary of changes occurring in the principal patterns.*

	Green-yellow-red blotched.	Yellow-green-red blotched.	Green-yellow spotted-red blotched.	Plants with laciniate leaves.	Green-red blotched.
Total number plants...............	337	41	198	68	90
Plants constant for green and yellow...	218	24	126	0	31
Percentage of constant plants.........	65	59	63	0	34
Changes in yellow:					
Increase:					
Frequency.....................	7	3	4	1	11
Ratio of frequency..............	9,630	2,730	9,900	13,600	1,630
Decrease:					
Frequency.....................	32	11	2	1	.....
Ratio of frequency..............	2,100	740	19,800	13,600	.....
Reversal:					
Frequency.....................	6	.....	.....	.....	.....
Ratio of frequency......	11,230	.....	.....	.....	.....
Total frequency....................	45	14	6	2	11
Ratio of frequency for all bud variations	1,490	590	6,600	6,800	1,630

limited to a branch could there by any degree of accuracy. Furthermore, fluctuations in number and size of the blotches of epidermal red, although frequent and somewhat persistent, were not recorded. As long as the pattern was blotched the plants were grouped together and changes to solid red or to no-red for considerable areas of a leaf were not considered as a bud variation unless a series of leaves showed that the change was sectorial for a stem. For this reason the data given in table 11 are summarized for fluctuations and bud variations involving yellow and green in patterns with red blotched epidermis.

The percentage of constant plants for yellow and green given in table 11 is derived by dividing the number of plants which were constant by the total grown of the pattern concerned. This gives an index of the constancy of a type, although it does not take into account the varying degrees of the fluctuations which appeared.

Judging from the data on changes in green and yellow, there appears to be no general correlation between the number of fluctuating plants and the number of cases of bud variations. In the *green-yellow-red blotched* group there were proportionally more than four times as many bud variations as in the group *green-yellow spotted-red blotched*, but the percentages of constant plants were nearly identical. Not one of the plants with laciniate leaves was constant for green and yellow, but only one case of bud variation occurred. In the *green-red blotched* group there was chance only for the appearance of yellow, and this change occurred in a relatively large number of cases, both in fluctuations and as bud variations.

A very marked contrast appears in a comparison of the two patterns *green-yellow-red blotched* and *yellow-green-red blotched*. Both have about the same proportions of green and yellow, except that the relative position is reversed. Both groups agree quite closely in the percentage of constant plants. In the latter, however, bud variations were 2.5 times as frequent. The position of the yellow in the center seemed to increase bud variations involving green and yellow over that in plants with the yellow at the border.

DISTRIBUTION OF BUD VARIATIONS AMONG DIFFERENT CLONES.

The wide range of variation both of fluctuations and of bud variations emphasized in the summaries already given was realized in a series of plants derived by vegetative propagation from two plants having the same color pattern. The records of pedigrees show that marked differences appeared among the various clones with respect to constancy and to the range and the frequency of bud variations.

This is shown quite clearly when the data regarding the main clones derived from plant 1 are grouped together as arranged in table 12.

TABLE 12.—*General summary of clones.*

Clone.	Total number plants.	Plants constant.	P. ct. of plants constant.	Number of bud variations.	Ratio of frequency.
11	211	132	62	49	1 : 860
12	192	87	45	21	1 : 1,830
13	138	75	54	4	1 : 6,900
14	155	80	51	18	1 : 1,720
117	91	54	59	31	1 : 590
111	34	29	85	4	1 : 1,700

The main clones 11 and 12 were derived from two branches of plant 1 which had the same color pattern. Although the branches were identical in appearance, the two progenies were quite different. 62 per cent of clone 11 were constant, while 45 per cent of clone 12 were constant; but in the more constant clone 11 there were proportionally

TABLE 13.—*Record of Clone 117.*

Generation and number of the plants.	Pattern of cutting.	Constant.	Fluctuating.	Spontaneous yellow.	Yellow-red blotched.	Green-yellow spotted-red blotched.	Green-red blotched.	Yellow-green-red blotched.	Green-solid red.	Green.	Green-yellow spotted.	Green-yellow.	Remarks.
1911–1912													
117........	2	×											Somewhat greener in winter.
1912–1914													
117, 1......	2			×			×	×					Each part distinct, but somewhat fluctuating.
1912–1913													
117, 11.....	2	×											Somewhat greener in winter.
117, 12.....	2	×											Do.
117, 13.....	6		×										Mixed during winter, both greener and yellower.
117, 14.....	2	×											Constant for pattern, somewhat greener in winter.
	5								×				Constant for loss of yellow.
117, 15.....	6				×	×							Well-defined sectorial bud variations.
117, 16.....	6	×											Slightly greener in winter.
117, 17.....	6	×											Do.
1913													
117, 111....	2						×						Each part constant.
117, 112....	2	×											Slightly more yellow as summer advanced.
117, 121....	2	×											Do.
117, 122....	2	×											Do.
117, 132....	6		×										Very irregular patterns.
117, 141....	2					×							Each part constant.
117, 142....	5	×											No trace of yellow.
117, 143....	2	×											Slight fluctuation in yellow.
	5	×											No trace of yellow.
117, 144....	5								×				Do.
117, 151....	6			×									Bud variation very marked; parts uniform.
117, 152....	6			×								×	Do.
117, 161....	6			×									Do.
117, 162....	6		×										Irregular patterns.
1913–1914													
117, 111, 1..	2	×											Quite uniform and constant.
	5	×											No trace of yellow.
117, 111, 2..	2					×							Constant except for bud variation.
117, 113....	2		×										Mixed patterns 2 and 4.
117, 122, 1..	2	×											Very uniform and constant.
117, 133....	6	×											Slightly greener in winter.
117, 134....	6	×											Do.
117, 135....	6						×						Constant except for bud variation.
117, 141, 1..	2	×											Slightly greener in winter.
117, 141, 2..	2	×											Much greener in winter.
117, 142, 1..	5			×									No trace of yellow except in two leaves late in summer.
117, 142, 2..	5			×									No trace of yellow except in one leaf late in summer.
117, 143, 1..	2	×											Slightly greener in winter.
117, 143, 2..	5	×											No trace of yellow.
117, 143, 3..	2	×											Very constant.
	5								×				Constant except for bud variation.
117, 144, 1..	5			×									No trace of yellow except in two leaves in late summer.
117, 151, 1..	1	×											Died in a few weeks.
117, 151, 2..	6						×						Bud variation sharply sectorial.
117, 151, 3..	1	×											Constantly very yellow.
	6	×											Slightly less yellow in winter.

TABLE 13.—*Record of clone 117*—Continued.

Generation and number of the plants.	Pattern of cutting.	Bud variations.									Remarks.
		Spontaneous yellow.	Yellow-red blotched.	Green-yellow spotted-red blotched.	Green-red blotched.	Yellow-green-red blotched.	Green-solid red.	Green.	Green-yellow spotted.	Green-yellow.	
1913–1914 Cont'd.											
117, 152, 1..	1										Very yellow at all times.
	6										Slightly less yellow in winter.
117, 152, 2..	1										Very yellow at all times.
	6										Slightly greener in winter.
117, 152, 3..	6										Do.
	14										Do.
117, 152, 5..	6										Do.
117, 152, 7..	1										Plant lived, but made poor growth.
117, 153....	5										Fluctuating to yellow spotted.
	6										Somewhat greener in winter.
117, 154....	6									×	Very constant except for bud variations.
117, 155....	6										Very constant.
117, 156....	6										Much greener in winter.
117, 157....	6										Do.
117, 161, 1..	1										Constantly very yellow, but greener in winter.
	6										Somewhat greener in winter.
117, 161, 2..	1										Do.
	6										Very uniform except for bud variation.
117, 161, 3..	1										Constantly very yellow.
	6										Very uniform except for bud variation.
117, 161, 4	1										This part soon died.
	6										Slightly greener in winter.
117, 161, 5..	1										Do.
	6										Mixed patterns, 4, 5, and 6.
117, 162, 1..	6										Slightly greener in winter.
117, 162, 2..	6										Do.
117, 162, 3..	6										Do.
117, 171....	6										Do.
117, 18.....	2										Lived only a few weeks.
	5										Soon died.
117, 19.....	5										No trace of yellow.
117, 1×1...	2										Very constant.
	5										Very constant; no trace of yellow.
117, 1×3...	2										Very constant except for bud variation.
117, 1×5...	2										
	5										Fluctuated to yellow spotted.
1914											
117, 111, 11.	5										No trace of yellow.
117, 111, 12.	2										Slightly more yellow as summer advanced.
117, 123, 1..	2										Do.
117, 133, 1..	6										Slightly more yellow, except for bud variation.
117, 136, 1..	6										Slight increase of yellow.
117, 136, 2..	6										Do.
117, 142, 11.	5										No trace of yellow except in two leaves.
117, 142, 12.											No trace of yellow except in one leaf.
117, 142, 21.											No trace of yellow except in three leaves (see text-fig.1).
117, 152, 31.											Slight increase of yellow.
117, 152, 32.											Do.
117, 152, 33.											Do.
117, 154, 1..											No trace of yellow except in branch with bud variation.
117, 154, 2..											Slight increase of yellow.
117, 156, 1..											Do.
117, 162, 11.											Do.

more than twice as many bud variations. Clone 13, which was derived from a branch that was *green-red blotched,* gave a progeny (of several patterns) of which 54 per cent were constant, but bud variations were very infrequent. Clone 14, with nearly the same percentage of constant plants, produced four times as many bud variations. This summary of the data by clones irrespective of patterns shows a general irregularity and lack of correlation between fluctuating variations and bud variations. The special interest, however, pertains to the clones 11 and 12, which show that two branches apparently identical may have quite different potentialities for constancy and for bud variations.

Even more marked differences than these developed among the various subclones. A study of pedigrees shows that in all patterns and in all main clones there were certain lines of progeny much more constant than many others. These could not be detected by any other than a pedigree method.

Clone 111 can be given as one of the most constant clones. Its members numbered 34. Four cases of bud variation appeared; 3 were a loss of yellow and 1 was a reversal of the position of the green and yellow occurring in one-half of a leaf only. These 4 plants were otherwise constant. Only 1 plant showed fluctuating variability, becoming quite uniform for *green-yellow spotted-red blotched.* All the bud variations involved changes in the green and yellow. There were no marked changes in the amount and distribution of epidermal red. As shown in table 12, the percentage of constant plants was 85 and the ratio of bud variations was 1 to 1,700. The clone was highly constant both in regard to fluctuations and bud variations.

On the other hand, the series of plants derived from plant 117 was 26 per cent lower in number of constant plants and gave nearly three times as many bud variations, yet plants 111 and 117 were both uniform and constant for the pattern *green-yellow-red blotched* and were apparently identical. Until the autumn of 1912, plant 1171 was the only one of the 17 plants grown in clone 11 that showed variation. It gave during the summer, by sectorial variation in the main axis, 6 branches with the position of the green and the yellow reversed. The plant was grown in a large pot during the winter and then grown out of doors during the following summer. In the second summer two more bud variations appeared on the part with *green-yellow-red blotched* foliage, but on branches quite separated. Both were sectorial; one was a loss of green, giving the *yellow-red blotched* pattern, and one was a loss of yellow, giving the *green-red blotched* pattern. The plant possessed for some time four patterns, each uniform for a certain part of the plant.

The record of pedigree for the progeny of the plant can be given as illustrating a clone in which bud variations occurred with a high ratio of frequency. In table 13 the plants are arranged in generations according to number. To trace the progeny or the ancestry of any plant, one should look in the generations following or preceding for the serial numbers. Certain plants were grown with two patterns as a chimera, and these are indicated by brackets, with a record for each pattern. In order to make the tabulation more compact, numbers are used to represent the different patterns, and these correspond with the numbers of the figures in the plate, as follows:

1 = *yellow-red blotched.*   5 = *green-red blotched.*
2 = *green-yellow-red blotched.*   6 = *yellow-green-red blotched.*
4 = *green-yellow spotted-red blotched.*   14 = *yellow-green.*

A survey of this series of plants shows that on the 91 plants 31 bud variations appeared, giving a ratio of frequency of 1 to 590 against 1 to 860 for the entire clone 11 and 1 to 1,700 for the sister clone 111.

A further analysis within this progeny shows that in several cases similar bud variations can be traced to a common ancestry. For example, 6 of the 9 cases of what has been called spontaneous development of yellow occurred during the summer of 1914 in plants descended from plant 117142 (see table 6), which itself was constant for loss of yellow during its period of growth. Another case of spontaneous development of yellow was in plant 1171441. In this clone, therefore, all cases of spontaneous development of yellow were in plants descended from plant 11714.

It is quite clear from such pedigrees that distinct differences in tendencies in regard to the degree of variation may exist among buds of branches bearing similar foliage.

ENVIRONMENTAL INFLUENCE.

Observations were made and pedigrees of plants examined to determine whether changes in ordinary environmental conditions influence fluctuations and bud variations.

To secure accurate data on the relative frequency of bud variations during summer and winter is hardly feasible. In general only about half of the summer plants were from cuttings made early in the spring. The others were grown in pots in a greenhouse during the winter and transplanted to the garden, where they grew during the summer, making larger plants with many more branches than were produced during the winter. Of the total number of 115 bud variations involving color, 9 appeared during the winter, which is fairly proportional to the relative number of branches that developed.

There was a strong tendency for plants having yellow to become greener in winter and yellower in summer, and also to become greener when severely pruned. At any time, however, during the winter, some plants of each pattern having yellow (for example *green-yellow-red blotched*) could be found with the pattern as pronounced as during the summer. During the winter of 1913–14, two plants of each of the following types, *green-yellow-red blotched, green-yellow spotted-red blotched*, and *green-yellow-solid red*, were grown in a greenhouse with northern exposure only, which gave scarcely any direct lighting. These plants were constant for the respective patterns. The green and yellow, however, were slightly less bright and intense.

During the summer of 1914 several plants of nearly all types of pattern containing yellow were grown in a greenhouse the glass of which was whitewashed to decrease the intensity of illumination. No appreciable differences could be noticed between the patterns of these and of plants grown out of doors. So far as I have observed, it does not seem that any of the color variations can be attributed to such factors as heat or degree of illumination.

Furthermore, the loss of yellow, loss of green, and gain and loss of red all occurred in single branches and in sections of branches (figs. 21 and 24). Frequently two quite different changes appeared on the same plant, which was then grown for some time with 2, 3, or even 4 quite distinct types of foliage. Cuttings were made so as to give plants with two types of foliage, as (a) *green* and *green-red blotched*, or (b) *green-yellow-red blotched* and *green-red blotched*, or (c) *green-yellow-red blotched* and *green-yellow-solid red*, etc. In all these cases branches with two types of foliage were submitted to as uniform conditions as possible; they grew on the same plant, were subject to the same degree of heat and illumination, and were supplied by the same root system. Under this test the different patterns were fully as constant as if grown on separate plants.

These facts indicate quite clearly that the marked and sudden variations and differences in expression of color concerned in the different patterns are not readily attributable to external environmental factors.

Flammarion (1898) used a variety of *Coleus* with a yellow, green, and red color pattern in testing the influence of light on pigmentation in plants. With red light he secured decrease of red pigmentation and a broader leaf; under influence of green light the red coloration mostly disappeared, and under blue light there was somewhat less red. A series grown out of doors under conditions of diffused light showed decrease of red coloring, while those under very dim light gave still less development of it. With decrease of red the center of the leaves became quite yellow. Evidently the red pigmentation of his variety was chiefly located in the epidermis. In marked contrast to these results it may be noted that the bud variations that I have reported give more marked changes than those induced by Flammarion and that these appear suddenly and in a sector of a bud in a manner that suggests internal readjustments rather than external environmental influence.

In respect to the laciniate leaf-shape and its periodic development, however, environmental influences seem to have some effect. As already indicated, conditions favoring rapid and vigorous growth lead to the development of entire leaves. In vegetative propagation, however, the periodic laciniate condition develops only in certain subclones. In other subclones it has not appeared. That is, the same conditions of environment and treatment do not lead to the appearance of the laciniate character in all plants, so it is hardly to be considered as purely environmental. The laciniate character has also developed in one plant in the manner of bud variations.

SEED PROGENY.

The data obtained from the seed progeny of my strains of *Coleus* have direct bearing on the nature and inheritance of the bud variations that appear and indicate that bud variations can give rise to as widely different forms as can be obtained among the various members of a hybrid progeny.

Selfed seed was obtained from a plant of the pattern *yellow-green-red blotched* and 22 plants were grown during the latter part of the summer of 1914. In respect to the development of green and yellow, there was every gradation between green with large yellow blotches irregularly distributed through the leaf-blade and pure green. In respect to the development of red in the epidermis, there were gradations from absence of red to a general distribution of large irregular blotches. As in the case of bud variations, the different types of epidermal red occurred independently of the degree of development of underlying green and yellow.

In regard to leaf-shape, there was every range of variation. Few plants could be classed as laciniate, but there was every gradation from shallow to deep lobing and from coarse to fine lobing. The leaves on any one plant were quite uniform. It should be noted that the laciniate character had not appeared in the particular subclone from which these seedlings were derived. It had appeared in sister subclones as described above. Seven plants possessed leaves quite like those of the parent type, both in respect to the cuneate base and the crenate margin. Five plants, however, had large leaves, some measuring 10 inches long, that were broadly obtuse at the base, with the cuneate character lacking. From plants possessing flat leaves with a smooth surface there was gradation to those with leaves much crinkled or folded.

At the same time 23 plants were grown from selfed seed of a plant which possessed the laciniate leaf-type as a fluctuating character. The plant itself also fluctuated in respect to the development of green and yellow as follows: during the winter of 1913–14, its leaves were strongly laciniate and devoid of yellow, and during the following summer the plant was quite yellow, becoming almost like the type *green-yellow-red blotched* (fig. 2) and every leaf was entire. In their development of green and yellow the plants ranged from pure yellow plants that died within a few weeks to those that were pure green. Two of the plants had much yellow in very irregular and mixed patterns. Eleven had no trace of yellow for several months, when a few yellow spots appeared in some of the leaves on several plants. In regard to epidermal red there were numerous types ranging from solid red to no red. Of those with red-blotched epidermis, some were uniformly finely blotched, while on others the blotches were large, with a single blotch sometimes covering one-fourth of a leaf. As to the shape of the leaves, the series showed the same range of variation exhibited by the seed progeny described above. There were but 5 that were strongly laciniate.

Besides the above plants which were grown to maturity, there are at the present date (February 8, 1915) 20 seedlings of a plant that had an epidermis of solid red, as shown in figure 8. The plants have from 2 to 3 pairs of leaves, but it is clear that only one of the seedlings has a solid red epidermis. Red-blotched types prevail and few of the seedlings show any yellow coloration.

Summarizing, it is plain that the plants grown from seed give wide variations. In respect to color patterns, there were numerous types which gave very complete gradations between extremes, especially in regard to epidermal red. Many of the types that had appeared as bud variations appeared also in the seed progenies, such as *yellow-red blotched, green-yellow spotted-red blotched, green-red blotched* (wide range of variation in respect to size of blotches), *green-yellow spotted,* and *green.* One of the new patterns could be described as *yellow-green blotched-solid red.* Another had the red blotches of the epidermis coalescing at certain points, making the red markings continuous but not solid (see fig. 29), so that the underlying green showed through in blotches.

In vegetative propagation leaf-shape remained very constant, the only exception being the clones that developed laciniate leaves. In the seed progeny of all plants tested, however, new types of entire leaves have appeared. The laciniate leaf characters and various intermediates between it and the typical entire leaf appeared in the first leaves of certain plants and remained as a rather constant character during the time the plants have been grown.

The wide variations appearing in the seed progeny indicate that this strain of *Coleus* is either of mixed parentage or that the processes concerned with production of color patterns and leaf-shape are themselves subject to wide variations. The variations in respect to color patterns were, however, no greater in range or extent than were those that appeared in bud variations, and the fluctuations from entire to laciniate leaf-shape gave extreme types of leaf-shape, with all grades of intermediates, on a single plant, as is quite well shown in plant 123153 of plate 4.

HISTORY OF COLEUS.

The early history of the cultivated varieties of *Coleus* shows that the original species utilized in cultivation and hybridization were few in number and relatively simple in respect to diversity of color patterns and leaf-shape.   The variations which appeared in vegetative and in seed propagation within a few years after introduction gave a wide range of variability, with greater extent than the extremes of the characters in the original species.   In general the variations reported in this paper are quite parallel with those appearing in the development of the numerous cultivated varieties, both in regard to the apparently spontaneous development of new patterns and to the reversion to parent or ancestral types.

It appears that the first variegated species of *Coleus* introduced into European cultivation was *Coleus blumei* Bentham.   The original description (Blume, 1826) under the name *Plectranthus scutellarioides* states that the leaves were spotted above with dark purple (folio supra maculis atropurpureis picta).   This plant was introduced into Holland in 1851 and the next year Planchon (1852) gave a brief description of it, accompanied by a colored plate.   It was soon introduced into England, and in 1853 a description with colored plate appeared in an English magazine (Hooker, 1853).   These two illustrations agree quite closely, although the latter shows the plant in somewhat a brighter green, with leaves with a somewhat more solid mass of central red.   The central part of the upper surfaces were dark purple or sanguineous, breaking into spots near the margin.   In the description it is stated that the leaves were entiré at the base, "which is gradually attenuated into a more or less elongated petiole," a character well shown in an outline drawing of a leaf.

There are no specimens of this species in the Bentham collections at the Kew Herbarium.   There is, however, a specimen in Herb. Hookerianum with a label in Sir William Hooker's handwriting, stating that the plant was grown at Kew Gardens and citing the description and plate in Botanical Magazine (Hooker, 1853).   This was evidently regarded by Hooker as a typical specimen.   The leaf-shape is identical with that of the illustration referred to and is, with the exception of figure 7, quite the same as the leaves shown in the plates illustrating this paper.   From the description and illustration the color pattern was nearly identical with my type *green-upper center solid red* (fig. 15).

Morren (1856) describes the variety *C. blumei pectinatus* as somewhat more richly colored, but differing chiefly in having the leaves deeply and doubly lobed.   The colored plate shows that the base of the leaf-blade was cuneate, as in the species.

Coleus verschaffeltii was first named by Lemoine (1861), who decided that it was distinct from *C. blumei*.   It appears (Witte, 1862) that this plant was introduced into Rotterdam in 1860 from Java.   Colored plates (Ill. Hort., **8**, pl. 293, and Flor. Mag., **2**, pl. 96) show that this species was richly and deeply colored with crimson on both surfaces and that the base of the leaf-blades were not cuneate but heart-shaped. Being richer in color than *C. blumei*, this plant attracted considerable attention as a foliage plant.   In 1864 (Gard. Chron., p. 506), a sport of this species called *marmoratus* with bright green patches in its leaves, was described.

Concerning *C. verschaffeltii*, Herincq (1865) remarks that the expression of red coloration fluctuates with light conditions and suggests that one might eliminate the red by keeping the plants in the shade. Later he (1866) notes fluctuations that are less due to environment, for he observes variation in the leaves of a single plant and states that no doubt selection of cuttings would give pure-green plants.   He mentions that he had seen a young plant having no trace of red coloration.

Two types of *Coleus* destined to play an important part in the development of horticultural varieties were introduced into England from New Caledonia by John G. Veitch.   Although briefly mentioned in 1866, they were first described and illustrated in 1867.   *C. gibsonii* (Verlot, 1866; Dombrain, 1867a) was of a dwarf bushy habit.   The leaves were large and "of a light-green color, distinctly veined and blotched with dark crimson-purple."   The plate clearly shows that the leaves were only slightly crenate and that the bases were broadly cordate. *C. veitchii* (Dombrain, 1867b) possessed leaves quite similar in shape, but with the entire central portion of the leaf of a deep chocolate color with the edges green.

These four species, *C. blumei*, *C. verschaffeltii*, *C. gibsonii*, and *C. veitchii*, were used as parents in the production of hybrids by F. Bause, in the employ of the Royal Horticultural Society of London.   12 hybrids, of which *C. verschaffeltii* was the seed parent, resulted the first year.   Rather extended descriptions of these are given by Thomas Moore (1868).   The F_1 progeny, even from the same parentage, were widely different, some resembling the seed parent, while others resembled the pollen parent.   In regard to leaf-shape, there were two groups, one with flat crenate leaves, as *C. veitchii*, and one with frilled-dentate leaves, as *C. verschaffeltii*.   In colors there were various shades of purple in solid colors, blotched areas, and in reticulations.   At the time this variation among the F_1 progeny aroused considerable interest.   One anonymous writer (Gard. Chron., **33**, 407) raises the question how several kinds of *Coleus* could originate from the same cross.   But one hybrid with *C. blumei* as a parent is reported.   This had frilled leaves and coloration much like that of *C. blumei*.   It was less deeply colored than the hybrids resulting for the other pollen parents.

There is no mention of yellow in any of these hybrids.   All were bi-colored, but with striking combinations of the green and various shades and amounts of purplish or red colorations.   They were sold at auction (Gard. Chron., **33**, 432) for the aggregate sum of £390. One hybrid brought 59 guineas.

The production of these valuable variegated *Coleus* varieties stimulated further hybridization work.   William Bull produced 18 different types (reported by Herincq, 1868), and in November of that year he advertised pedigreed seed from 20 crosses involving 14 varieties grown that year (Advertisement, Gard. Chron., **33**, 1232).

Meanwhile, at the gardens of the Royal Horticultural Society, a new series of hybrids were produced much finer than those of the previous year.   The parentage of these interesting hybrids is not fully given, but it is stated by Moore (1869) that certain of the hybrids of the previous year were crossed with *C. blumei* itself.   In this second lot of hybrids, yellow coloration appeared as a new or spontaneous development.   Eight (Gard. Chron., **33**, 1210) possessed distinct yellow, forming in some cases a golden margin.   Two (Prince Arthur and Princess Beatrice) are described as having a yellowish ground-color or golden green.   Most of them had a yellow and green ground-color overlaid with shades of purple or crimson red.   The most brilliant of the series was named Queen Victoria, a colored plate of which appeared as a frontispiece in the Florist and Pomologist (volume for 1869).   This plate shows that the ground-color was mostly yellow, overlaid by an epidermal red, appearing crimson over the yellow and entirely covering the upper surface of the leaves except at the margin. None of the series possessed frilled leaves.   The leaves of all were flat, with crenate teeth somewhat deeply cut.

Although of much more remarkable variegation than the hybrids of the previous year, 9 of these new coleuses brought but 65 guineas.

While the yellow element in the variegation appeared strongly in this second lot of hybrids, it should be noted that in the year 1867 (Gard. Chron., **33**, 460) a golden *Coleus* arose as a bud-sport from *C. blumei*. It is described as like *C. blumei*, but with the green exchanged for a decided yellow tint.   The sport appeared in one-half of a single leaf. The bud at its base was propagated and gave the new variety.   It does not appear that this sport was used in the hybridization work that produced the golden coleuses.

During 1868 and 1869, the various horticultural publications mention by name no less than 54 new varieties of *Coleus*.   For several years thereafter few varieties were mentioned, but in 1878 (Gartenflora, p. 50) 13 forms not previously mentioned are listed.   The next year this journal (pp. 341–346) states that breeding of *Coleus* had been carried on in Germany, speaks of new forms that arose, and prints an uncolored plate illustrating 4 types.   New types were also credited to Bull (Rev. Hort. Belg., **5**: 49; Gard. Chron., **45**: 748).

At the exposition in Paris in 1879, Morlet exhibited varieties described by André (1879) as far surpassing all previous varieties.   These had enormous foliage, with remarkable combinations of shades of carmine, yellow, and green. 20 types are mentioned by name and 12 of these are described.

Pynaert (1881) states that about 250 varieties of *Coleus* had been put upon the market. He notes that it is difficult to establish a nomenclature for the various *Coleus* varieties. He describes and illustrates in color a variety called Reine des Belges which arose from seed of the variety Duchess of Edinburgh, a variety illustrated in the Floral Magazine in 1874. The leaves of the Duchess of Edinburgh possessed a yellow border; the Reine des Belges had the yellow in the center of the leaves. The relative position of the green and the yellow was therefore reversed, which is the difference between types *green-yellow-red blotched* and *yellow-green-red blotched* already reported in this paper.

From the evidence at hand it is clear that a large number of types of *Coleus* have been produced. Probably the same type or nearly identical types have been given different trade names. With the exception of the first hybrids produced by Bause, there is almost no record of the parentage of most types. The plants attracted attention solely on account of their variegated foliage, and for a time were more extensively used as bedding and foliage plants than they are at the present time.

From the standpoint of genetics, it is suggestive that such wide variation appeared in the cultivation and hybridization of the 4 species already discussed, although it should be noted that there is the possibility that other species were concerned in the parentage of some of the varieties now in cultivation. The parent species possessed a green background, or at least were without pronounced yellow. The epidermis especially was more or less colored with purple or red in *blumei*, *verschaffeltii*, and *veitchii*, while in *gibsonii* the purplish coloring was largely confined to the veins.

In the varieties derived from these, yellow appeared as a pronounced part of the coloration. Some types were largely yellow, others were pale yellow, and others were entirely green. In many the yellow was localized at the border, but in others it was at the center and in others the yellow blotches were well distributed. These variations in yellow and green were combined with variations in amount and quality of epidermal and internal (chiefly in veins) development of purplish and red tints.

The historical evidence indicates that the form of *Coleus* used in the experiments here reported is derived from *C. blumei*. In respect to the cuneate base of the leaf-blades and the marginal character of the first pair of lateral veins almost making the petiole attenuate, the leaves of all types but the laciniate agree almost exactly with that of the figure for *C. blumei* (Bot. Mag., 1853). This character differentiated sharply *C. blumei* from *C. verschaffeltii*, *C. veitchii*, and *C. gibsonii* or any other species introduced into cultivation, and seems to indicate clearly that the type in question is derived more or less directly from *C. blumei*. The flowers agree with the description of *C. blumei*.

Coleus blumei, as stated above, was already in cultivation in Java when described by Blume. Blume (1826) suggests that the plant named by him *Plectranthus laciniatus* may have been simply a variety, as it seems to have differed largely in having laciniate leaves. Blume's original description speaks of the leaves of *C. blumei* as spotted above, but the colored plates appearing in 1852 and 1853 show the greater part of the upper surface of a solid purplish color. *Coleus blumei* was introduced into the German gardens under the name *Plectranthus concolor* var. *picta* (Gartenflora, 1853, **2**: 220). Only in one of Bause's first lot of hybrids was *C. blumei* concerned, but this species was, it is stated, the seed parent of the second lot of hybrids. If this hybrid was used in further hybridization work there is no record. It was the least brilliantly colored, possessed no yellow, and sold for the sum of 5 guineas, which was the lowest sum paid for any one of the 12 hybrids (Gard. Chron., **33**, 432). I have been unable to find further mention of this hybrid, which was named *reevesii*.

C. blumei produced in 1868 a bud sport with the green changed to a decided yellow tint. Through propagation this gave rise to the variety *telfordi aurea* (Gard. Chron., **33**, 460).

André (1880) illustrates and describes 4 new varieties, which he attributes to *C. blumei*. Apparently all have, however, leaves with cordate bases and not at all cuneate, as is the case in *C. blumei*, which makes his determination of doubtful validity. The same is true of the type Gloire de Dijon, described and figured by Rodigas (1888). Rodigas later (1892) notes the wide variability obtained from the seed progeny of what was considered as *C. blumei* from Chile. The 4 derived types illustrated possess, however, strongly cordate leaves which make the identity with *C. blumei* doubtful.

At the present time it does not appear that any pure strains of *C. blumei*, *C. gibsonii*, or *C. veitchii* are in cultivation. The strain used in these experiments agrees most closely in regard to leaf-shape with the original *C. blumei*, but the variability of the seed progeny seems to indicate that it is not a pure strain. The writer inquired about and observed all types of *Coleus* available at numerous botanical gardens and nurseries during a 6 weeks' trip to Germany, Holland, and England during the summer of 1914. Only one plant, a plant observed at the Royal Botanic Gardens at Regent's Park, London, was seen which had the *blumei* type of leaf.

Coleus verschaffeltii is, however, quite generally in cultivation at the present time and agrees quite closely with the type first described.

I am especially indebted to Mr. F. J. Chittenden, of the Royal Horticultural Society, for securing a statement (a letter to Mr. Chittenden) from Mr. B. Wynne, in which he states that he spent 3 months of his time (as a student) at Chiswick in 1866, under Bause, in the propagating department.   He was quite familiar with the methods used in the development of the hybrid coleuses and states that "it fell to my lot to convey the first half dozen coleuses sold to Sleven's rooms." Mr. Wynne refers to the description (see reference above) of the hybrids as adequate.   In regard to the present existence of these hybrid types, Mr. Wynne gives the following statement:

"I am unable to say whether there is more than one of the Chiswick-raised set of *Coleus* in existence now, but I very much doubt it.   It is very interesting, however, that at least one of them has survived and is still grown for Covent Garden and possibly for other markets.   This is the variety originally named Queen Victoria, which received a first-class certificate (R. H. S.) in the autumn of 1868, and was bought at the second sale by the now extinct firm of John and Charles Lee, of Hammersmith.   I do not think it is now known in the market by its original name, but it is the well-known variety with chocolate leaves and golden segments and I have no doubt about its identity with our old Chiswick plant."

It should be noted that Wynne's description of this form as "chocolate leaves with golden segments" does not agree with the colored plate in the Florist and Pomologist for 1869.   The latter represents the pattern as solid crimson in the center, with well-defined yellow border.

While *Coleus* is now regarded with less favor than it formerly received, there are many types in cultivation exhibiting great range of color patterns and leaf character.   At Erfurt, Germany, during the summer of 1914, the writer saw large collections of *Coleus* grown for seed for the trade.   In the Ernst Benary collections they were chiefly of large-leaved, small-leaved, and fringed-leaved types.   In the large-leaved plants the yellow, if present, was usually in the center, a condition which was true without exception for the plants with small leaves. The fringed-leaved types possessed most curious doubly cut proliferations about the margin.   Of the entire collection hardly any two plants were alike as to color distribution.   There were, however, fewer classes in regard to leaf-shape.   One of the newer types was a dwarf with fringed leaves derived by selection.

At the greenhouses of Haage and Schmidt there were 2 rather definite types with laciniate leaves.   One is *salicifolia*, with narrow, slender, quite irregularly-lobed leaves.   Another is *quercifolia*, with broad leaves coarsely cut with round-tipped segments.   Their large-leaved types, entire and fringed, showed great diversity of color patterns. The plants are grown for seed, and in producing stock for seed parents selections are made with special reference to leaf-shape and general habit of growth.   In general no selection is made with reference to color patterns and each type exhibits wide variation in this respect.

From the history of *Coleus* it seems quite clear that the numerous and diverse varieties have arisen from few species.   These varieties exhibit many characteristics of coloration and leaf-shape that were not possessed by any of the parents; for example, there is no evidence that a yellow element of variegation was present in any of the parent species. In leaf-shape also, many new types have arisen. Hybridization and rather intense and artificial cultivation have been associated with the development of such diversity from plants comparatively simple.

Considering the various types of *Coleus* as a whole, we may note that the wide range of variability which they exhibit is in large measure realized in the bud variations that have appeared from the single types here reported. The *green, green-yellow, yellow-red blotched,* and *green-solid red* are extreme forms that are in degree and quality counterparts of the extremes seen in the different varieties. In the development of the laciniate character of the leaf-shape in which the leaves on a single plant fluctuate from entire to extremely laciniate, the counterparts of cut-leaved *quercifolia* and *salicifolia* types are in large degree realized.

In seed progeny and in bud variations the single strain of *Coleus* investigated has in the 3 years of observation shown the same types of variation that have developed in the entire series of cultivated varieties derived by both seminal and vegetative variations.

DISCUSSION.

The appearance and subsequent behavior of bud variations in *Coleus* present numerous analogies to various phenomena of variation exhibited by members of a seed progeny of hybrid origin.

In respect to the definiteness of the characters contrasted, green and yellow, red and non-red, the color patterns arising by bud variations are as different, at least in their extremes, as one could expect in the members of a seed progeny even of hybrid origin. This is especially noticeable in comparison with the seed progenies of *Coleus* itself. The solid-red epidermis and the no-red epidermis represent two extremes fully commensurate with the so-called presence and absence of a character and the bud variations giving these were fully as different as the types arising in seed progeny. The intermediates *red blotched* and *solid red upper center* are pattern characters equally distinct both in manner of appearance and in vegetative constancy. The same is true of the extremes of development of both green and yellow.

*P*lants of the same pattern in a single line of descent, both in the same generation and in successive generations, frequently produced the same type of variation independently, a behavior quite analogous to the segregations that reappear in each generation of a hybrid line or in the successive generations of the progeny of a mutating plant. Many of these variations show a return to a parental pattern, just as a recessive parental quality reappears as a result of segregation in hybrid progeny. There is much in such reappearance of patterns that is quite identical with the phenomena of alternative inheritance.

Two plants identical in appearance and derived from adjacent branches on the same plant may give quite different progenies in successive generations. One line may be very constant and uniform, the other may give numerous bud variations of wide range. This is a familiar phenomenon in hybrid seed progenies where certain of the plants of any generation, although apparently identical, give quite different progenies. In Mendel's experiments with *Pisum*, for example, although the F_2 yellow peas of the cross green $\times$ yellow were similar, they gave different progenies. Some produced only yellow peas, while the progeny of others gave both yellow and green.

In the strains of *Coleus* studied by the writer, certain types of bud variation occur more frequently than others. Loss of yellow was more frequent than loss of green, and loss or decrease of red was more frequent than increase of red. In the entire series of plants derived by vegetative propagation there was decided predominance of *green* over *yellow*, of *red blotched* and *no red* over *solid red*, of the *blumei* character of leaf-base, and of the shallow or crenate lobing of the leaves. These same characters show marked tendencies for dominace among the members of the seed progeny.

Such phenomena of variation appearing in hybridization experiments are usually considered as due to segregation and recombination of hereditary units during the processes of self- or cross-fertilization. Bud variations in vegetatively propagated plants are, of course, independent of such recombinations.

That bud variations are generally due to a complete loss during cell division is not substantiated by the results here reported or by the bulk of other experimental work. In the majority of cases the character concerned does not breed true. Mendelian students have interpreted this to mean that such bud variations are produced by a loss of only one factor of a diploid pair, giving heterozygocity. This illustrates the tendency of the Mendelian interpretation to assign the numerous cases of fluctuation in characters to heterozygocity rather than to fluctuations in a factor or to irregular mutational changes spontaneous in the organism.

In regard to the range of expression in a single plant the laciniate leaf-shape is a more striking character even than the color patterns. It arose, as already noted, in 13 individuals obtained by vegetative propagation, but these were all derived from a few plants of the next preceding generation. This character was inherited through vegetative propagation by all plants grown but one, but the leaves on each individual plant varied from deeply laciniate to fully entire. Plants raised from seed gave all types from extremely laciniate to fully entire, the particular type appearing in the first leaves that developed and remaining quite constant for all leaves developed in the 6 months that the plants have been grown. The special point of interest is that a single individual of the laciniate group passed through a series of fluctuations, giving all grades of leaf-shape from entire to fully laciniate. The range in a single plant is greatly more marked than the differences between the *Urtica* hybrids (*C*orrens, 1905, 1912), in which the serrated type of leaf was dominant. In the hybrids of the normal and the laciniate types of *Chelidonium majus* (de Vries, 1900) there seems to be no published data regarding the range of variation in the F_2 generation. Of the hybrids between palm-leaf and fern-leaf types of *Primula sinensis*, Bateson (1909) states that "dominance is usually complete," but that he has seen two strains with intermediate leaf-shape. Gregory (1911) states that "the palmate character is dominant, though a slight difference can sometimes be recognized between pure and heterozygous palmate types." *C*rosses between an ivy-leaf (a palmate shape with margins crenate) with the fern-leaf gave the normal palmate leaf as an F_1 hybrid. The F_2 generation exhibited a wide variation, which Gregory groups into 4 classes and assumes that shape of the entire leaf and crenation of the margin are two independent characters.

On the basis of character of lobing of climax leaves, Shull (1911) distinguishes four biotypes in *Bursa bursa-pastoris*. The view that hybrid forms always segregate out into only these types is somewhat in doubt, for Shull (p. 9) finds that a plant classed as *simplex* gave unexpectedly a mixed progeny with defective ratio, so that it is clear that the assumed hereditary "gene" became less potent. Hus (1914) distinguished in a culture of *Capsella bursa-pastoris* 4 forms different from those of Shull and added another factor which he considers determines the narrow character of early leaves in certain forms.

In none of these studies have the individuals of the F_2 exhibited greater variation among themselves than have single plants of *Coleus* with the laciniate leaf, nor have the individual parent plants been more distinct and uniform as a plant than the plants derived from seed progeny of *Coleus*, and grown for a period of several months. Furthermore, there has been evidently no attempt to select persistently intermediate types for modified potency of characters. Furthermore, emphasis has not been laid to selection of variations in a seed progeny of a single individual or in a line propagated vegetatively. The more intensive Mendelian studies, such as those by Shull, Hus, and Gregory, indicate that the character of leaf-shape is complex and that selection studies along the lines indicated may reveal further data on variations in the potency of characters.

The historical evidence and the studies of seed progeny reported above indicate that the strain of *Coleus* studied is most certainly of mixed parentage. If this strain had been studied solely in its seed progeny the variations obtained would be attributed by many modern geneticists, I venture to imagine, to chance combinations of hereditary units. Yet, as has been fully reported above, these variations are analogous to and even identical in nearly all cases with those arising by bud variation.

At this point we may note that modern genetics have furnished no evidence as to the real nature of the characters considered. What we may call the genetical or breeding value of characters has alone been emphasized. Characters have been considered solely in regard to their unity of expression in successive generations of plants of selfed or hybrid origin. Special emphasis has been placed on the reappearance of characters and upon their phylogenetic significance. From this standpoint we may further consider specifically the characters concerned in these studies of *Coleus*.

Considering first the color characters, we note that the pattern of the parent plants was a mosaic of *green, yellow, red* (or *blue*), and *non-red* cells. In the mature leaf the cells are apparently qualitatively different, and furthermore, the color differences between the various cells are identical with the color differences between entire leaves and branches derived by bud variation. The contrast between pure yellow and pure green leaves in the bud sports and on plants of seed origin is the same that exists between green and yellow cells that may be adjacent in the same leaf. Likewise, the differences between *solid-red* epidermis and the *non-red* epidermis is the same that exists between *red* and *non-red* cells in the epidermis of a single leaf having a *red-blotched* pattern. In the blotched patterns the cells of the epidermis are either red or non-red. The number of similar cells that are adjacent to each other determine the size of the blotches.

The facts regarding development and distribution of the colors raise sharply the question whether the possibility for development of red, yellow, and green is possessed by every cell that is formed or whether these possibilities have been sorted out by qualitative divisions in a Weismannian sense. A sharp distinction should be made between a character that is metidentical and one that can only belong to a group of cells. In Detto's (1907) sense, power to produce green, yellow, and red seems to be metidentical; that is, this ability may be strictly a property of all cells, while the differeut patterns appearing so strikingly as characters of entire leaves can not in any sense be metidentical.

Most important evidence regarding the development of leaves with *red-blotched* patterns was obtained by the cytological studies conducted by Mr. E. G. Arzberger. In the early stages of leaves that later will become *red-blotched*, all the epidermal cells possess red pigmentation, forming a complete periclinal layer of red cells, and in this respect the early stages of leaves with *solid-red* and *red-blotched* patterns are alike. In *solid-red* patterns all cells continue to possess red sap color, while in *blotched* types the red disappears in certain cells. The evidence is clear in these cases that all the epidermal cells arise from cells having red cell sap and in this respect are potentiaHy alike.

The differences in patterns in respect to red are largely those of quantity; the total of red cell sap in a leaf is comparatively large in solid-red patterns and much less in blotched patterns, of which there is every grade to the no-red epidermal patterns. Differences in distribution are also involved. There is the tendency for red to be in the epidermis rather than in the subepidermal tissues. Usually it is in both upper and lower surface, but in one pattern (fig. 10) the red is almost entirely massed in the upper epidermis and in the center of the leaf. In the blotched types the number, size, and degree of coalescing of the blotches present every gradation from a finely blotched condition to a solid red.

While the total amount of red and its distribution in the epidermis determines quite definite and constant patterns, it should be said that no plant has been obtained, either in seed or vegetative propagation, that was entirely devoid of red. Subepidermal cells may also possess red cell sap; red sap is quite pronounced in veins of leaves (see figs. 13 and 14), and varying amounts can be seen in stems. In some plants, especially those of seed origin, no trace of red can be seen except at the nodes, and here the amount for different plants varies from faint rings to quite a definite band or zone of red cells extending through the stem. Also the stems of plants, especially those with blotched patterns, may have large irregular streaks of red-colored tissue and we may say are internally blotched. In general, plants with a decided nodal zone of red were those with epidermis free of coloration, and on the other hand, plants with solid epidermal red or heavily blotched epidermal red had as a rule slight localization of red in zones. There was also the marked localization of red in *upper center* of leaves (see fig. 10), with almost complete absence of red below, although in many leaves of this type small blotches of red were evident on the under side (see fig. 10a).

All these conditions indicate that the possibility of producing and possessing red cell sap is a specific property of all cells and that the distribution giving localization at nodes, in streaked areas in stems, and in subepidermal tissues, and in the epidermis either as a uniform red or a blotched red are dependent on two fundamental conditions: (*a*) total amount of pigmentation, and (*b*) the appearance of it in certain centers of concentration. The facts as described for *Coleus* seem to indicate the such conditions are determined largely by intercellular relations.

Chemical studies, in general, show that differences in quality, quantity, and distribution of pigments in flowers and leaves are correlated with changes in quality, amount, or distribution of any one of such substances as chromogens (glucosides, phenols, tannic acid, etc.), oxydases, enzymes, oxygen, etc.

That marked changes in color quite comparable to those I have described for *Coleus* may result from slight chemical changes is well shown by the chemical studies of various members of the genus *Monarda*. The results of these studies are fully summarized by Wakeman (1911). The isolation and determination of yellow and red pigments and a study of their chemical relatives have given rise to the quinhydrone hypothesis of plant pigmentation. It is considered that the plant oxidizes thymol or carvacrol to a series of oxidation products of yellow, orange, and red colors, but all closely related to each other. Of these, hydrothymoquinone, thymoquinone, and dihydroxythymoquinone have been definitely isolated and determined. These in turn have the capacity of adding phenols yielding highly colored phenoquinones and quinhydrones.

Wakeman (1911, p. 111) states:

"Taking into consideration only those compounds that have been isolated or whose presence has been indicated in the monardas thus far, the number of possible pigments becomes truly bewildering."

Furthermore, some of the pigments are phenolic in character and can combine with metallic constituents of the plants, giving rise to different shades of the original pigment.

The highly colored red and purple pigments of the stems and leaves, and the yellow and purplish pigments of the flowers in *Monarda fistulosa* are thus quite definitely identified as mixtures of quinhydrones which are shown to be direct products of the plant. The important point is that while the pigments are themselves of highly complex chemical substances, the changes which produce marked differences in color are very slight.

It is also significant that marked differences may exist between the parts of a single flower. The simplest of color patterns in *Antirrhinum* treated by Wheldale (1914, p. 110) illustrates this condition. Flowers of the yellow-flowered variety possess a pale yellow pigment in the tube of the corolla, a deep yellow pigment in the lips, and a still deeper patch on the palate. In attempting to harmonize these varied conditions with a Mendelian factorial analysis, the power to produce these pigments in spatial relation is assigned to a factor Y. In the "ivory" flower a pale pigment is found in the tube and lips and a yellow pigment only in the epidermis of the palate. Power to produce pale pigment quite generally in the flower, and to inhibit the formation of deep yellow everywhere but in the epidermis over the palate is assigned to a single factor, I. The conditions of color distribution are in themselves color indicators that different processes occur in different parts of the same flower, giving different kinds of substances in different amounts and with different distribution.

Studies of oxydase reactions in different tissues indicate that intercellular relations are of much importance in determining the distribution of pigments. By means of micro-chemical tests Keeble and Armstrong (1912) obtained evidence that the distribution of oxydase in various types of *Primula* is closely correlated with the development of anthocyanin. In general they find the oxydase most abundant in epidermal and in vascular tissue. The extent of oxydase distribution differs much in different varieties. They make the highly interesting observation that action of chloroplasts seems to act as an inhibitor of oxidase formation or of the production of chromogen. Varied types of color pattern in flowers, and even anthocyanin development in stems and leaves, is, according to these investigations, closely related to distribution of oxydases and chromogens.

It is of further interest to note that patterns resulting from such qualitative and quantitative reactions which depend in considerable degree on flow of substances in a plant may be quite uniform not only in the flowers of a plant, but among the various members of its seed progeny. Such phenomena have led to the assumption that patterns are represented as such in germ-cells by hereditary units. When, however, hybridization occurs between varieties having even the simplest of patterns, the F_2 generation more often than otherwise presents a most remarkable range of types. Usually, this sort of diversity results in crosses between varieties, especially when the color patterns of flowers and leaves are concerned and gives results that could not be predicted with any degree of accuracy.

Riddle (1909), in a comprehensive survey of the chemical and physiological facts regarding the origin and nature of melanin pigments, points out (p. 323) that "a single chromogen acted upon by a single enzyme usually produces several colors depending upon the degree of oxidation involved." He gives conclusive evidence that "the power to oxidize tyrosin compounds is not dependent primarily upon germinal segregation, but rather upon active tissues, relations, and conditions," and that local conditions, especially in pathological cases, determine the production of melanin. In general Riddle shows the inadequacy of a strict interpretation of color inheritance in animals on the basis of unit factors and gametic purity.

In the light of all the chemical studies on pigmentation, it seems clear that qualitative reactions are concerned which involve the production, flow, and assembling of substances through the relations and interactions between cells. That these interactions should be so widely readjusted in a hybrid progeny resulting in such varied expression of color in quantity, quality, and distribution is suggestive that fundamental readjustments may occur more readily with characters that are dependent on cellular interactions than with those that are strictly metidentical. As already pointed out, the variations in *Coleus* propagated vegetatively give numerous patterns differing widely in regard to quantity and distribution of the pigments concerned and present the same sort of phenomena of readjustment seen in seed progeny of hybrids.

In questions relating to the development of color patterns, the Liesegang precipitation phenomena, especially as developed by Gebhardt (1912) and Küster (1913), seem to me most illuminating. By the various phenomena associated with rhythmic precipitation and crystallization of mineral solutions in gelatin plates, Gebhardt was able to produce simple and multiple eye, line, and flaked patterns strikingly similar to various markings in butterfly wings. By varying the substances used, modifying the amounts, the distribution, the degree of concentration, and providing for interaction between areas of different concentration, a wide range of markings were produced. Gebhardt points out that such physical and chemical phenomena indicate that the distribution of pigment even in the intricate markings of butterfly wings may be due to an epigenetic regulation of the quality and quantity of such substances as chromogen, oxydase, and oxygen and the reciprocal influences of different centers of action. It is pointed out that the cell boundaries and especially the position of veins may determine the distribution of the substances involved and determine the relative locations of centers of action as well as areas and centers of no action.

Küster (1913) extended the study of Liesegang precipitation phenomena to the effects produced in capillary tubes, obtaining various types of banded precipitation patterns which, as he emphasizes, suggest that similar chemical and physical processes may determine many types of variegation in both monocots and dicots.

In *Coleus* the development of patterns is considered by the writer to be due largely to cellular and tissue interactions influencing general and metidentical qualities with results quite analogous to the Liesegang phenomena. Changes involving red are on this basis rather simple cases of readjustment influencing the total amount of pigment produced and the distribution in centers or areas. The ability to produce the different chemical substances concerned with the final development of the red pigmentation is assumed to be a general property or potentiality of all *Coleus* cells. The assembling of all the products necessary for the final stages in its development, however, are determined by the amounts produced and their flow to centers of activity and interaction. The development of red, especially in the subepidermal tissues, indicates that this is the case and suggests strongly that, as shown by Overton (1899), changes in the amounts of red pigmentation may be closely related to changes in the sugar-content of the sap. In *Coleus*, however, it is clear that such changes arise quite spontaneously in the cells and tissues.

In respect to the development of green and yellow in particular cells, the processes seem to be antagonistic. Plastids are present in both green and yellow cells, but in yellow cells they are fewer in number, smaller in size, and somewhat distorted in shape. The green and yellow cells are subepidermal, extending from upper to lower epidermis. In a pure-green leaf all these cells remain green. In the most extreme cases of yellow development nearly all the cells fail in the production of chlorophyl. The different patterns result from variations in the relative number of green and yellow cells and in the grouping of the cells of like color. In some a green field is blotched with island-like areas of yellow cells, in others the central area of green is bordered by an irregular band of yellow, and again the yellow may be situated in the center with a green band at the border.

In a leaf with blotched or banded green-yellow patterns the intermingled areas of green and yellow cells indicate quite clearly that both types of cells are derived from the same cells in the growing-points. In the development of leaves it can be observed that while the yellow areas are in evidence when the young leaves unfold, the yellow seems tinged with green, and that as the leaf grows the yellow becomes more intense. Furthermore, the yellow bleaches until in old leaves the yellow areas change to a pale yellow or white, while the green areas are still bright. As a leaf dies the green areas become pale greenish yellow. These observations indicate that many of the cells which later become yellow are actually green at first.

The fluctuations that appear substantiate this view. A plant with yellow-bordered leaves may produce, especially in winter, new leaves entirely green, and thus possess for some time old leaves of *green-yellow-red blotched* pattern and younger leaves of *green-red blotched* pattern.

Later the leaves formed may show increased amounts of yellow, until by midsummer the plant is uniform for the *green-yellow-red blotched* pattern.　Figures 14 and 14*a* show winter and summer conditions quite general for the leaves of the *yellow-green* pattern, which indicates that cells may be either green or yellow.　Furthermore, all degrees of variation can appear as bud variations affecting segments of a bud.

Such conditions indicate that all the cells are potentially green. If this be accepted, a further point is raised regarding the source of the influences leading to loss of green and to development of yellow in its place.　In respect to the final action in the cells, change from green to yellow is itself apparently a local phenomenon in that local action of plastids is affected.

The fluctuations and variations in the extent and position of the green and yellow tissues, however, indicate that here, as in the development of red, certain intercellular influences are operating.　The configurations of the yellow and green areas, as well as that of the red, are in marked degree bounded by veins.　For the green and yellow this is most strikingly shown in figures 12 and 14.　For the red similar conditions are seen in any of the blotched patterns, as, for example, figures 2 and 5.　This also indicates that the flow of substances giving different centers of distribution and concentration is the important factor in the production of patterns.

In respect to the extent and degree of the variations, it has already been pointed out that no plant in my cultures of *Coleus* has been obtained without some red coloration in some part of the plant.　In regard to yellow, however, there were frequently fluctuations and bud variations giving branches with no yellow.　The loss of yellow appears to be complete in a manner that suggests loss of hereditary qualities through segregations, but even in constant selection of pure-green stock for vegetative propagation about half of the offspring show return to patterns containing yellow both by fluctuations and by bud variations, with, also, cases of marked spontaneous appearance of yellow in a few or in single leaves.　To say that the power to produce yellow has been latent is to say that the conditions causing its development can arise in an apprently spontaneous manner.

While the evidence indicates that the ability to produce green, yellow, and red is a metidentical property of the cells, it is equally clear that these metidentical properties do not exist as units.　They are subject to interaction between cells.　They are more or less permanently modified either by the intercellular relations or by spontaneous intracellular changes.　The records of pedigree show very clearly that tendencies to give vegetative progeny of different degrees of constancy and variation arise or exist in sister branches that are apparently quite identical.　Such tendencies detected by the pedigreed cultures have already been mentioned.　The production of branches which give clones of plants with decreased amounts of red indicates a specific decrease in power to develop red and the results of pedigree culture indicate that continued selection in this direction would give lines with only slight amounts of red or perhaps that are entirely free of all red.

The metidentical characters of green, red, or yellow are themselves fluctuating not only in expression but in inheritance through cell lineage. They do not appear to be independent. I have never yet obtained a plant by seed or by vegetative propagation that did not possess some degree of red coloration, and as yet no strain has been isolated that was pure for loss of yellow.

On the whole, vegetative propagation of any new type that arises gives a progeny that exhibits a rather marked degree of constancy with fluctuations and variations about a new mode. Selections for pure green do not give a progeny of pure-green plants, but do give a greater number of green plants than does any other pattern. Selections for decreased red or for increased red likewise give clones with this tendency prevailing. In other words, readjustments of the processes concerned with total production of pigments and their distribution tend strongly to occur in growing-points or to so affect them as to secure a certain degree of permanency.

The assumptions of de Vries already noted in the introduction seem to apply quite adequately to the behavior of the metidentical characters as far as expression in individual cells is concerned, but does not fully explain the phenomena of pattern changes as well as the conception of a further influence of intercellular relations, modifications of which may in time affect more or less permanently the expression of metidentical qualities. Any Mendelian conception of pattern factors that are units in heredity is quite inadequate, as is also such a conception for even the metidentical characters.

This analysis of the nature of variegation and the significance of bud variation in *Coleus* has a direct bearing on the nature of certain other types of variegation.

One of the most clearly marked types of variegation is that of the infectious chlorosis. The best known cases are those of tobacco (Beijerinck, 1899; Woods, 1899) and *Abutilon* (Baur, 1904 and 1906; Lindemuth 1897, 1899, 1901, 1905, and 1907). All investigations agree that in these types the variegation is not transmitted to seed progeny. The searching investigations of Beijerinck and Baur lead to the conclusion that a living fluid or virus carries the contagion. In the case of tobacco the infection is readily accomplished by various agencies. In *Abutilon* grafting is necessary, and by this means the variegation has been transmitted to numerous species of *Abutilon* and related genera. Similar types of infectious variegation exist in other groups of plants, as *Fraxinus, Jasminum, Liburnum,* and *Ligustrum.*

In the infectious variegation of the type seen in *Abutilon striatum thompsonii* the pattern is a mottled one, with irregular yellowish areas mingled with the green. The amount of yellow varies considerably, especially according to intensity of illumination. In *Abutilon megapotamicum variegatum* especially. the distribution on a single plant is very irregular (see Reid, 1914). Only a few blotches may appear on a leaf and often entire leaves or all leaves on an entire branch may be pure green. Baur (1906a) found green branches on *Abutilon striatum thompsonii* and was able fully to establish that they were immune.

Immune branches arise on a plant as bud variations, but the leaves differ from those with variegation only in having all cells immune, for in the latter a part of the cells perhaps remain immune. It may be noted that blotched variegation in this case either results from irregular immunity or to irregular distribution of virus, and hence emphasizes the intercellular relations concerned with distribution.

Whatever the nature of the "virus" may be, it is fully demonstrated by Baur that it is a product of the diseased cells of the old leaves and is transported to young leaves in which certain areas of cells succumb to the influence while others do not. This immunity exhibited by some cells, however, may suddenly extend to entire leaves or to all leaves on a branch. As to the flow of the "virus," Baur (1906a) found that it could pass through tissues of immune strains of *Abutilon arboreum*, causing infection to non-immune parts beyond, but that such infection was not produced if immune tissue of *Lavatera arborea* intervened.

It should be noted that this type of variegation can not, as far as we know now, be distinguished by appearance from other types of mottled variegation. Its infectious nature and the failure to transmit to seed progeny are the characteristics of these cases. As noted above, Shull (1914) suggests that certain yellow-flecked types of variegation giving very irregular transmission to seed progeny may be of such infectious nature that it can be carried in some of the germ-cells. It may well be that in many cases of variegation (especially of the blotched types), the disturbing cause producing loss of chlorophyl may be quite similar in nature to that of the vigorously infectious types. In the latter it is quite clear that the production of a "virus" in variegated leaves and its flow to young leaves does occur.

Differences in the extent of influence of such a "virus" may give in some types an apparent inheritance through seed progeny. In the infectious types we know nothing definite concerning the appearance of variegation in the plant first showing it. It appears (Reid, 1914) that many, if not all, of the abutilons with infectious variegation arose through grafting with one original strain, *Abutilon striatum thompsonii*. The presence of infectious variegation in such widely differing genera as *Nicotiana*, *Fraxinus*, *Abutilon*, and *Ligustrum* indicates that the condition should arise spontaneously. Frequent and almost continued spontaneous development of a less vigorous virus may be very common, and many cases of variegation, even of those that are apparently seed constant, may be due to such a condition.

Lindemuth (1905) determined that the variegation in *Coleus* was not infectious like that of *Abutilon*. Just what types he used is not clear from the data given. Conclusive evidence regarding this point has not been obtained by the writer. Thus far grafting experiments between *green-yellow* and *pure green* types have shown no cases of development of yellow, and the writer has assumed that the variegation is not at least vigorously infectious.

We may also note that spontaneous loss of ability to produce green in a part of the cells of the growing-point may result in a chimeral variegation such as Baur (1909) reports in white-margined types of *Pelargonium zonale*. His anatomical studies showed that in growing-points the white-colored tissues may lie over the green, forming histogenic layers and giving remarkable permanency of the periclinal chimera in vegetative propagation.

In maintaining this relation white cells give rise to white cells and green to green, but mechanical readjustments in the growing-points may give branches with quite different distribution of the two kinds of cells. Branches may thus arise with sectorical distribution of green and white, with only white or only green cells, or even with reversed positions (Baur, 1909; Stout, 1913). These readjustments give no new qualities to cells, nor do they appear to involve changes of either kind to the character of the other. Yet the occurrence of numerous types of variegation with this chimeral relationship indicates that such spontaneous loss is not infrequent.

Certain types of bud variation in *Coleus* present features quite similar to the readjustments that appear in *Pelargonium*, and raise the question whether there is possibility of spatial readjustments of distinct tissue elements. The sudden and apparently complete loss of epidermal red suggests that this layer may exist as peripheral in a more or less chimeral relationship, but the development of red and non-red in the adjacent cells of the epidermis of a single leaf indicates clearly that these differences can arise within cells of the same immediate progeny. If for any reason a part of the epidermal cells fail to develop red, the red might be absent in the entire epidermis for the same reason.

The loss of yellow giving pure green might seem to be due to the exclusion of yellow cells. Also, the sharp contrast between patterns of *green-yellow* (figs. 2, 8, and 12) and pattern of *yellow-green* (figs. 6, 11, and 14) suggest the possibility of a spatial readjustment in the growing-point of two distinct cell elements. But certain plants in all these patterns have fluctuated, getting greener to a *green-yellow blotched* pattern during winter and returning to the type pattern in the next summer, and often exhibiting at one time in a single row of leaves various gradations between the extremes.   The extreme yellow types show some green areas of tissue.   The extreme green type, which appears to have no yellow, gives numerous fluctuations, and cases of well-marked spontaneous development of yellow.   The distribution of green and yellow does not in any pattern show anatomically a chimeral distribution, as both colors are much intermingled in the subepidermal tissue.

That a certain degree of chimeral relationship exists in certain patterns of *Coleus* is evident.   In the patterns with solid-red epidermis the epidermal layer is rather specialized in respect to concentration of red cell sap.   These apparent chimeral relationships in *Coleus* are due to intercellular development of patterns rather than to specific and qualitative differences in cells as such.

Numerous cases of variegation are induced by environmental conditions.   *C*ramer (1907, chap. xi) summarizes cases of the influence of parasites, of soil conditions, light, and temperature in producing certain types of chlorosis and variegation.   These are, we may say, direct reactions to external conditions, which in most cases are quite apparent.   At first thought this class may seem quite distinct from what are considered as true hereditary types, but the difference is chiefly one of degree, for there are few types of cases of variegation that do not fluctuate in response to certain environmental conditions. The infectious types of variegation fluctuate very much according to degree of illumination and may entirely disappear from a plant if it is kept in darkness for sufficient length of time.   The types with green and white as periclinal chimeras show, perhaps, least fluctuation in regard to environmental influence.

In this respect what I have called fluctuations in *Coleus* are of interest.   Fluctuations in amount of red present in blotched types is constantly occurring.   One plant, at the present writing, has some branches with leaves sparsely blotched, as in figure 5, others with the red blotches strongly coalescing, as in figure 29, and still others with nearly a solid epidermal red.   These fluctuations do not seem related to external environmental conditions.   In the amounts of green and yellow, the various patterns possessing these two elements showed a strong tendency to be more green in winter and more yellow in summer.   The degree of fluctuation was, however, not uniform for the different subclones, for the several plants of a single generation, or even for all the branches of a single plant.   A few plants maintained a maximum of yellow in their leaves throughout the winter.   Still it is very clear that many cells in the leaves are green in winter which would have been yellow had the particular leaves developed during summer.

In certain plants variegation appears periodically.   These types are, perhaps, in a strict sense, to be classed with the preceding, but differ at least in showing marked periodicity which may well be due in large measure to inner conditions. *C*ramer (1907, p. 128) summarizes numerous cases of those showing marked differences in behavior. *C*ertain varieties of *Quercus* have pure-green foliage each spring, but later produce variegated leaves. *Ulmus scabra* var. *viminalis* is yellow during midsummer, but pure green at close of the summer. *Linaria biennis* is pure green during the first year of growth, but variegated in the second.

These phenomena illustrate again that cells of the same lineage may fluctuate in development from a maximum of green to various degrees of loss of chlorophyl, with often development of yellow coloration, and that such phenomena may reappear with marked constancy in progeny.

Aside from these classes there is a wide range of types, including many cases in which the variegation reappears more or less generally and constantly in the individual plants and in the variety. The variegation appears to be inherited, at least in a certain degree.

In the propagation of variegated plants, much general data has been obtained regarding the degree of constancy both in vegetation and seed propagation. *C*ramer (1907, p. 129) gives a summary of numerous cases where the variegation disappears in certain methods of vegetative propagation. In *Cornus mascula variegata* (T. M., Gard. *C*hron., **32**: 952), root-cuttings give pure-green individuals, but plants from layers retain the variegation. Such cases, however, should be investigated with special regard to the nature of the variegation. In many of the cases noted, especially those in *Pelargonium*, the variegation may be chimeral, and when root-cuttings are made the green cells have greater power of regeneration. The inconstancy exhibited in vegetative propagation, however, is no greater or more marked than that which develops on a single individual. Exact evidence of the inheritance and constancy of variegation through pedigreed vegetative progenies seems to be lacking. While in some varieties the variegation appears to be quite constant, in others it is widely variable.

The Mendelian studies that have been made of the seed progeny of variegated plants and of the bud variations which involve changes of pattern show likewise a wide range of behavior. Hybridization studies involving variegation, as has been pointed out in the introduction, indicate clearly the wide range of variability, and what, from the Mendelian viewpoint of unit characters or unit factors, is most erratic inheritance. These studies have contributed interesting and valuable data on the sort of variations one may espect in hybridization studies of this kind, but they indicate very clearly that the assumed factors are themselves fluctuating. In these studies the variability of plants used as seed parents has not been determined by vegetative propagation. This is not possible in all cases, but whenever it is possible the emphasis should be placed on this line of investigation if one is to speak with certainty regarding the nature of the inheritance.

If we turn our attention to the various phenomena associated with the laciniate character of leaves and petals, we find the same degree of irregularity and diversity of origin, expression, and inheritance that is exhibited by variegations. Cramer (1907, chap. 21) devotes a chapter to an excellent survey of the facts regarding the behavior of the character. It is interesting to note that Cramer observes that varieties quite constant in seed progeny are likewise very constant in vegetative propagation. There are numerous cases known of the spontaneous development of a laciniate type from one with entire leaves, both as seed mutations and as vegetative mutations; likewise of return to the types with entire leaves. While some cases are quite constant, others are widely fluctuating, even exhibiting a marked periodicity.

The behavior of this character in *Coleus* is most striking in its variability of expression and of its inheritance as a periodic variation through vegetative propagation.

Fundamentally, the processes involved in the development of leaf-shape are quite different from those involved in the production of pigments such as green, yellow, and red. The shape of the leaf in general depends on the rate, number, and regularity of cell divisions in the different planes of growth. It would seem that a general and quite uniform series of cell-divisions would give a leaf of more regular outline, and that if the cell-divisions in the growing leaf occur irregularly, giving, so to speak, points or lines of more rapid growth somewhat analogous to apical growth, with a more or less multiple dichotomy, then cut, lobed, or laciniate leaves would result. The shape of the leaf, it would seem, is determined by intercellular relations concerned with the manner of cell-divisions.

A comparison of the variability that develops in vegetative propagation with that occurring in the seed progeny reveals some essential differences between the inheritance of the characters involved. While the range of variation is quite the same, there is a marked difference in what we may call the intensity of variation. In vegetative propagation the degree of the intensity was low, with reference to the frequency of the appearance of new color patterns or to the development of the laciniate character. A large number of plants were grown by vegetative propagation. The bud variations were comparatively infrequent, occurring something like once for every 10 plants grown, giving in the course of 6 generations the different types of pattern described. In a seed progeny, however, practically the entire range of variation which appeared in vegetative propagation was seen in a single seed generation comprising 50 plants. The processes concerned with reduction and fertilization increased the intensity of variation and brought out in a single progeny of no great number the full extent of variations.

Color patterns, which are intercellular characters and in a sense vegetative types, are inherited through vegetative propagation in marked degree, while in seed propagation there is no evidence that they are inherited as such. The intercellular relations involving amounts and distribution of pigments are widely and suddenly disorganized and readjusted during the processes concerned with seed formation. On the other hand, the metidentical characters green, yellow, and red are quite uniformly transmitted in both sorts of cell-divisions.

The behavior of the laciniate character is significant in this connection. After this character made its appearance in a plant it was a constant feature in the development of all plants (but one) derived by vegetative propagation. The seed progeny, however, exhibited wide differences, ranging from deeply laciniate through all degrees or grades to the entire type of leaf. Furthermore, the laciniate leaf appeared in all seed progenies thus far grown, even when derived from plants of a line in which the character had never appeared. The processes of reduction and self-fertilization in a *Coleus* plant seem to bring out latent possibilities for various developments of leaf-shapes.

It remains to be seen if any of the types appearing in seed progeny are more constant in vegetative or in seed propagation than are the similar types that develop by bud variation. Already several bud variations have appeared in plants of seed progeny, indicating vegetative changes in the processes concerned with pigment and pattern formation. The colors involved in the variegation of *Coleus* represent every type of coloration (green, yellow, white, and red or blue) concerned with the variegation and coloration of plants. There is no evidence that the essential nature of these characters in *Coleus* differs from that of the characters concerned with variegation and pigmentation in corn (Emerson, 1914), in *Mirabilis* (Correns, 1909), in *Melandrium* (Shull, 1914), in *Pelargonium* (Baur, 1909), in *Antirrhinum* (Baur, 1910).

The explanations here given regarding the spontaneous variability of the characters concerned in the development of pigments and of the changes in intercellular and intertissue relations influencing development of color patterns in *Coleus* apply equally well to such cases as those just noted. The evidence indicates that the characters in question, and most especially the pattern characters, are not represented by units or factors, unless these assumed factors are to be considered in a general sense as temporary conditions descriptive of types of development and not as particular localized units of germ-plasm, which is the conception that gave the Mendelian interpretation its definiteness and simplicity.

The knowledge of the nature and the heredity of color characteristics will be advanced more by studies of the natural variability of the characters involved and by chemical and physical investigations of the processes concerned in the formation and distribution of such substances as melanin, flavone, and phenol compounds rather than by further elaboration of complicated formulæ involving multiple factors that attempt to explain fluctuations, inherited variations, and cases of increased variability that appear in hybrid seed progenies.

SUMMARY.

(1) A single variety of *Coleus* propagated vegetatively by cuttings in two main clones has shown (*a*) gradual fluctuations and (*b*) sudden mutations, giving a total of 16 distinct and characteristically different color patterns.

(2) (*a*) A total of 15 patterns (see diagram 2) arose by sudden mutation affecting a part of a leaf or a branch, or a series of associated leaves or branches; (*b*) 6 of these 15 patterns (figs. 2, 4, 5, 10, 13, and 13*a*) also appeared among the fluctuating variations; (*c*) one type of color pattern (fig. 15) has thus far appeared only as a fluctuating variation.

(3) (*a*) Six (diagram 2, and figs. 1, 4, 5, 6, 8, and 12) of the 15 color patterns arose directly from the parent type by sudden bud variation. One of these 6 (fig. 4) also appeared as a fluctuating variation.   (*b*) Five of these 6 types (figs. 4, 5, 6, 8, and 12) propagated by cuttings showed further fluctuations and bud variations, giving (*a*) the parent type (fig. 2), (*b*) 4 of the 6 types already directly derived (figs. 1, 2, 5, and 12), and (*c*) 8 new types (figs. 9, 10, 11, 13, 14, and 16, also 8*a*, 13*a*, not illustrated).

(4) The variations in the development of color patterns mentioned above involve (*a*) increase and decrease of green and yellow, (*b*) increase and decrease of red pigmentation, (*c*) reversals of the relative positions of the green and yellow by which a type with green center and yellow border (fig. 2) gave one with yellow center and green border (fig. 6), and (*d*) changes in the distribution of the red pigmentation especially giving concentration in the epidermis of the upper surface of the leaves.

(5) Progeny of 11 types of color pattern have been grown through from 2 to 6 generations, as follows:

Type of figure	1	2	4	5	6	8	9	10	12	13	14
Number of generations	2	6	6	6	6	5	2	2	2	2	2
Total number of plants	11	337	198	90	41	54	8	7	7	7	4

Some of these types have shown themselves more constant than the parent type (fig. 2), others were less constant.   All varied about a new mode and all would be considered good horticultural races.

(6) The relative constancy of color-pattern types derived by the accumulation of fluctuating variations was tested in two cases: Type of figure 2 (see clone 14 of table 2), 3 generations, total, 45 plants; type of figure 4 (see clone 13 of table 3) 3 generations, total, 79 plants. In both cases the constancy of the progeny showed no essential difference from that of the same types obtained by sudden bud variations.

(7) Variations in leaf form, even more striking than changes in color patterns, from entire to deeply laciniate-leaved forms, appeared in 13 instances as fluctuations affecting an entire plant and in one case (during the winter of 1914–14) as a bud variation.   The striking

feature of this variation is the marked tendency to the production of laciniate leaves during winter or after a period of particular drought during summer.

(8) The behavior of the laciniate character was observed in three subclones (see table 4) of 4, 2, and 2 generations, totaling 68 plants. In 67 of these plants the laciniate type of leaf appeared as a definite character, with a marked tendency to a periodic summer and winter fluctuation.

(9) Plants with the laciniate leaf character also showed wide fluctuations in regard to the degree of green or yellow coloration. When grown for a period of a year from cuttings made in autumn, the leaves were as a rule entire and slightly yellow in autumn, deeply laciniate and pure green in winter, and entire and very yellow during the following summer.

(10) In sexual reproduction all the principal types of variegation and leaf-shape appear at once in an F_1 generation, with also numerous types that were intermediate and fluctuating. The extremes of variation are no greater than those obtained in vegetative propagation, although some new types of entire leaves were thus obtained.

(11) Between any two types numerous intermediates arose, showing that we have here no evidence of the somatic segregation of invariable pattern factors.

(12) In the bud variations, decrease of red occurred with about twice the frequency as did increase of red; likewise decrease of yellow occurred about twice as often as the increase of yellow, indicating a definite tendency for variations in the direction of the increase of green and the decrease of red. These facts are doubtless due to fundamental relations between the chemical compounds involved.

(13) The types of color changes involving (a) green and yellow and (b) red and non-red occurred entirely independently of each other.

(14) The types produced by bud variations are the equivalents of the "Kleinarten" or the "biotypes" commonly occurring in cultivated species propagated by seed.

(15) Selection within clones is effective in securing progenies of new types with as high degrees of constancy as is possessed by ordinary cultivated races.

(16) The results indicate that slight variations arising either as sudden mutations or as gradual fluctuations can perpetuate themselves.

(17) The green, yellow, red, and non-red colorations in *Coleus* can best be characterized as metidentical characters; that is, they are the same in the cells as in the tissue and their appearance is possible in the development of any cell.

(18) The distribution of the colors giving pattern characters are properties of groups of cells and tissues as such. Pattern characters are probably due entirely to tissue and cellular interactions.

(19) The explanation suggested by the production of patterns in colloids by the Liesegang precipitation phenomena, especially as applied by Gebhardt to the markings of butterfly wings and by Küster to the development of many types of variegation in plants, seems to apply to the production of color patterns in *Coleus*. On this view color changes may be considered as due to the formation of different diffusion centers for the development and concentration of pigments.

(20) Bud variations in *Coleus* are (*a*) common; (*b*) give numerous different types which may be vegetatively quite constant from the first or can be made so by selection; (*c*) show development of certain types more commonly than others; (*d*) produce reversions to parental types; (*e*) give development of different degrees of variability among sister clones; (*f*) exhibit spontaneous changes in the fundamental color characters (metidentical) and in the cellular and tissue processes resulting in color patterns.

(21) The results show that in *Coleus* asexual and sexual reproduction are not fundamentally different in respect to the extent and range of variation.

A. B. Stout,
Director of the Laboratories.

New York Botanical Garden,
New York City, February 10, 1915.

BIBLIOGRAPHY.

ANDRÉ, ED. 1879. Les nouveaux Coleus. Ill. Hort., 26: 160.
———. 1880. *Coleus blumei*, nov. var. *hortenses*. Ill. Hort., 27: 52-53; pl. 377.
BATESON, WILLIAM. 1902. Mendel's principles of heredity.
———. 1909. Mendel's principles of heredity.
———. 1914. Address of the President of the British Association for the Advancement of Science. Science, II, 40: 319-333.
BATESON, WILLIAM, and R. C. PUNNETT. 1905. A suggestion as to the nature of the "walnut" comb in fowls. Proc. Cambridge Phil. Soc., 13: 165-168.
BAUR, ERWIN. 1904. Zur Aetiologie der infektiösen Panachierung. Ber. Deutsch. Bot. Gesell., 22: 453-460.
———. 1906a. Weitere Mitteilungen über die infektiöse Chlorose der Malvaceen und über einige analoge Erscheinungen bei *Ligustrum* und *Laburnum*. Ber. Deutsch. Bot. Gesell., 24: 416-428.
———. 1906b. Ueber die infektiöse Chlorose der Malvaceen. Sitz.-ber. Preuss. Akad. Wiss., 1906: 11-29.
———. 1909. Das Wesen und die Erblichkeitsverhältnisse der "Varietates albomarginatæ Hort." von *Pelargonium zonale*. Zeit. Ind. Abs. Vererbs., 1: 330-351.
———. 1910. Untersuchungen über die Vererbung von Chromatophorenmerkmalen bei *Malandrium*, *Antirrhinum*, und *Aquilegia*. Zeit. Ind. Abs. Vererbs., 4: 81-102.
BEIJERINCK, M. W. 1899. De l'existence d'un principe contagieux vivant fluide, agent de la nielle des feuilles de tabac. Arch. Néerl., Ser. II, 3: 164-186.
BLUME, C. L. 1826. *Plectranthus scutellarioides* Bl. Bijdragen tot de Flora van Nederlandsch Indië, p. 837.
CASTLE, W. E. 1912. The inconstancy of unit characters. Am. Nat., 46: 352-362.
CORRENS, C. 1905. Ueber Vererbungsgesetze.
———. 1912. Die neuen Vererbungsgesetze.
———. 1909a. Vererbungsversuche mit blass (gelb) grünen und bunt blättrigen Sippen bei *Mirabilis jalapa*, *Urtica pilulifera*, und *Lunaria annua*. Zeit. Ind. Abs. Vererbs., 1: 291-329.
———. 1909b. Zur Kenntnis der Rolle von Kern und Plasma bei der Vererbung. Zeit. Ind. Abs. Vererbs., 2: 331-340.
CRAMER, P. J. S. 1907. Kritische Uebersucht der bekannten Fälle von Knospenvariation.
DARWIN, CHARLES. 1868. Variation of animals and plants under domestication. 2 vols.
DETTO, CARL. 1907. Die Erklarbarkeit der Ontogenese durch materielle Anlagen. Biol. Centralb., 27: 80-95, 106-112, 142-160, 162-173.
DE VRIES, HUGO. 1889. Intracellulare Panagenesis.
———. 1900. Das Spaltungsgesetz der Bastarde. Ber. Deutsch. Bot. Gesell., 18: 83-90.
———. 1901. Die Mutationstheorie.
———. 1913. Gruppenweise Artbildung.
DOMBRAIN, H. H. 1867a. *Coleus gibsonii*. Floral Mag., 6: pl. 338 (with description).
———. 1867b. *Coleus veitchii*. Floral Mag., 6: pl. 345 (with description).
EAST, E. M. 1908. Suggestions concerning certain bud variations. Plant World, 11: 77.
———. 1910a. The transmission of variations in the potato in asexual reproduction. Contrib. Lab. Genetics, Bussey Inst. Harvard Univ., No. III. Biennial Report Conn. Exp. Sta., 1909-10, 119-161.
———. 1910b. A Mendelian interpretation of variation that is apparently continuous. Am. Nat., 44: 65-82.
EMERSON, R. A. 1914. The inheritance of a recurring somatic variation in variegated ears of maize. Research Bull. Exp. Sta. Neb., No. 4.
FLAMMARION, CAMILLE. 1898. Physical and meteorological researches, principally on solar rays, made at the Station of Agricultural Climatology at the Observatory of Juvisy. U S Dept. Agr., Exp. Sta. Record., 10: 103-114, pl. 1.
GEBHARDT, F. A. M. W. 1912. Die Hauptzüge der Pigment Verteilung im Schmetterlingsflügel in Lichte der hiesegangschen Niederschläge in Kolloiden. Verh. Deutsch. Zool. Gesell., 22: 179-204.
GOODSPEED, T. H. 1912. Quantitative studies of inheritance in *Nicotiana* hybrids. Univ. of Calif. Publ., Bot., 5: 87-168.
GREGORY, R. P. 1911. Experiments with *Primula sinensis*. Jour. Genetics, 1: 73-132; pl. 30-32.

HERINCQ, F. 1865. Choix de plantes a feuillage rouge. Hort. Français, 7: 308–311. Also same title and article in La Belg. Hort., 1865, 15: 293–295.

———. 1866. Production de variêtês par le bouturage. Hort. Français, 8: 237–240.

———. 1868. Coleus nouveaux des Anglais. Hort. Français, 10: 271–274.

HOOKER, W. J. 1853. Coleus blumei. Curtis's Bot. Mag., III, 9: pl. 4754 (with description).

HUS, HENRI. 1914. The origin of × Capsella bursa-pastoris arachnoidea. Am. Nat., 48: 193–235.

KEEBLE, FREDERICK, and E. F. ARMSTRONG. 1912. The role of oxydases in the formation of the anthocyan pigments of plants. Jour. Genetics, 2: 277–311, pl. 19.

KÜSTER, ERNST. 1913. Ueber Zonenbildung in Kolloidalen Medien. Beiträge zur entwicklungs mechanischen Anatomie der Pflanzen.

LEMOINE, CH. 1861. Coleus verschaffeltii. Ill. Hort., 8: pl. 293 (with description).

LINDEMUTH, H. 1897. Vorläufige Mitteilungen von Veredelungs versuchen innerhalb der Malvaceen und Solanaceen. Gartenflora, 46: 1–6.

———. 1899. Kataibelia vitifolia (Willd.) mit goldgelb marmorierten Blättern. Gartenflora, 48: 431–434.

———. 1901. Impfversuche an Malvaceen. Gartenflora, 50: 8–11.

———. 1905. Ueber verschiedene Arten der Panaschüre, deren Uebertragbarkeit durch Transplantation und Samenbeständigkeit. Gartenflora, 54: 125–128.

———. 1907. Studien über die sogenannte Panaschüre und über einige begleitende Erscheinungen. Landw. Jahrb., 36: 807–862.

MENDEL, GREGOR. 1865. Versuche über Pflanzen-Hybriden. Verhandlungen des Naturforschendenvereines in Brün, 4: (Abhandlungen) 1–47. (Translation in "Mendel's Principles of Heredity" (third impression), 1913, 335–379.)

MOORE, THOMAS. 1868. New hybrids of Coleus. Gard. Chron., 33: 376–377.

———. 1869. Coleus Queen Victoria. Florist and Pomologist, 1869, 1-2, pl.

MORGAN, T. H. 1913. Factors and unit characters in Mendelian heredity. Am. Nat., 47: 5–16.

MORREN, EDOUARD. 1856. Le Coleus blumei, variété pectinatus. Belg. Hort., 6: 99, pl.

NILSSON-EHLE, H. 1908. Nôgat om nuvarande principer vid hösthveteförädlingen på Svalöf. Sveriges Utsädesförenings Tidskrift.

———. 1909. Kreuzungsuntersuchungen an Hafer und Weizen. Lands Univ. Årssk., Teil. 1.

OVERTON, E. 1899. Beobachtungen und Versuche über das Auftreten von rothem Zellsaft bei Pflanzen. Jahrb. Wiss. Bot., 33: 171–231.

PLANCHON, J. E. 1852. Coleus blumei. Fl. Serres, 8: 141.

PUNNETT, R. C. 1911. Mendelism. Third edition.

PYNAERT, ED. 1881. Coleus reine des Belges. Ill. Hort., 28: 102–104. pl. 425.

REID, KATHERINE W. 1914. Variegated abutilons. Jour. N. Y. Bot. Gard., 15: 207–213.

RIDDLE, OSCAR. 1909. Our knowledge of melanin color formation and its bearing on Mendelian description of heredity. Biol. Bull., 16: 316–351.

RODIGAS, ÉM. Coleus blumei Benth. var. nov. Coleus Gloire de Dijon. Ill. Hort., 35: 31, pl. 46.

———. 1892. Coleus blumei Benth. varietates novæ. Ill. Hort., 39: 111–112, pl. 164.

SHULL, G. H. 1908. The pedigree culture: its aims and methods. Plant World, 11: 21–28, 55–64.

———. 1911. Defective inheritance ratios in Bursa hybrids. Verh. Nat. Ver Brunn, 49: 1–12.

———. 1912. "Phenotype" and "clone." Science, II, 35: 182, 183.

———. 1914. Ueber die Vererbung der Blattfarbe bei Melandrium. Ber. Deutsch. Bot. Gesell., 31: (40)–(80).

STOUT, A. B. 1913. A case of bud-variation in Pelargonium. Bull. Torrey Club, 40: 367–372, pl. 20.

VERLOT, J. B. 1866. Plantes nouvelles, rares ou peu connues. Revue Hort., 37: 279–280.

WAKEMAN, NELLIE. 1911. The Monardas. Bull. Univ. Wis. Sci., 4: 81–128.

WEBBER, H. J. 1903. New horticultural and agricultural terms. Science, II, 18: 501–503.

WHELDALE, M. 1914. Our present knowledge of the chemistry of the Mendelian factors for flower-colour. Jour. Genetics, 4: 109–129. pl. 7.

WITTE, H. 1862. Coleus verschaffeltii Lem. et son introduction en Europe. Ann. Hort. Bot., 5: 125–128.

WOODS, A. F. 1899. The destruction of chlorophyll by oxidizing enzymes. Centralb. Bakt. II, Abt. 5: 745–754.

EXPLANATION OF FIGURES IN PLATES 1, 2, 3, AND 4.

The paintings here reproduced were made by Miss Mary Eaton, artist of the New York Botanical Garden. All figures are reduced to about three-fourths natural size.

PLATE 1.

FIG. 1. A typical leaf of the pattern *yellow-red blotched* taken from plant 1171613 on February 21, 1914.

2. The pattern *green-yellow-red blotched*. Taken from plant 121 on April 15, 1912. The yellow border is somewhat irregular and is not so fully developed as in summer.

4. Leaf classed as *green-yellow spotted-red blotched*. Painted on January 30, 1913. Shows rather few scattered yellow spots and a few rather large epidermal blotches.

5. Pattern *green-red blotched* from plant 131, taken April 15, 1912. Shows the complete loss of yellow as it occurred in the first bud variation observed and as it frequently appears in plants having yellow.

5a. Young leaf of the pattern *green-red blotched*.

6. The pattern *yellow-green-red blotched*. Taken from plant 1171 on November 2, 1912. Shows the type which appeared on a plant with the pattern of figure 2 by a reversal of the relative positions of the green and yellow.

7. A good example of the laciniate type of leaf. Taken on January 31, 1912, from a plant quite identical to plant 123153 shown in plate 4.

8. The *green-yellow-solid red* type. Differs from figure 2 in having a solid red instead of a red blotched epidermis. Taken on January 4, 1913, from plant 32 and shows the typical development of the yellow, which was very constant and uniform on the plant during the winter.

9. A typical leaf of the *green-solid red* type. This differs from figure 8 only in having no yellow in the subepidermal tissues and from figure 5 in possessing a solid red epidermis.

9a. A young leaf of the pattern *green-solid red*. The red completely covers the leaf and is of the same intensity as in a mature leaf.

10. This figure shows the upper surface of a leaf of the pattern *green-yellow-solid red upper center*. Figure 10a is of the under surface of the same leaf. The distribution of the green and yellow is as in figure 2. The red pigmentation is almost entirely massed in the epidermis of the upper surface. There are a few small areas of red in the lower epidermis. Painted on December 9, 1913.

11. The type *yellow-green-solid red* painted on October 26, 1914. This type developed from type of figure 8 by a reversal of the relative positions of the green and yellow, a change which also gave figure 6 from figure 2.

PLATE 2.

FIG. 12. Type *green-yellow*, painted May 17, 1914. The pattern is slightly irregular in this leaf, with a yellow segment extending to the midrib. This pattern was derived from type of figure 2 by a loss of epidermal red. There are a few streaks of red in the vascular tissues.

13. The *green* pattern showing some epidermal red in the vascular tissues and a pale and diffuse shade of yellow. Painted on June 5, 1914.

13, a, b, c, d. Four successive leaves of a plant of the same line of descent as the plant of figure 13. Painted on February 2, 1915. The series shows an increase of red in the subepidermal tissues as the leaf matures, with the maximum development in figure d. Comparison with figure 13 shows the increase of red in a line of descent by gradual fluctuation. The distribution of red indicates a relation involving vascular tissues.

14. Pattern *yellow-green*. This was derived from type 6 by a loss of epidermal red. The streaks of red are subepidermal. Painted on February 11, 1914, showing the decreased development of yellow frequent in this type during winter.

14a. The typical summer condition of the type *yellow-green*. Painted on May 29, 1914. The yellow areas are bounded in marked degree by veins and there is a much less development of green cells in the central area.

PLATE 2—Continued.

FIG. 15. A leaf with no yellow and with almost no red on the under surface. A type quite like that described for the original *Coleus blumei*. This type appeared thus far only as a fluctuating variation as a winter condition of a few plants of the type of figure 10. At the time this was painted, January 24, 1914, the oldest leaves of the plant were typical for type 10 and the younger leaves were uniform for the coloration here shown. The under surface of this leaf had only a few red blotches similar to those of figure 10a.

17. A leaf with irregular pattern, developed from type of figure 2.

18. One of the leaves of the plant shown in figure 21, showing the absence of yellow in half of a leaf.

19. Leaf painted on January 18, 1914, showing rather marked increase of yellow in the type of figure 2 during winter. Painting was made after the yellow areas had begun to turn to white.

20. Painted on July 5, 1913. Typical of the most extreme fluctuation of type figure 2 in regard to increase of yellow.

PLATE 3.

FIG. 21. Young plant 12514, grown from a bud showing sectorial loss of yellow, by bud variation in half of the bud. Photographed and painted on November 9, 1912.

22. One of the leaves in a branch showing sectorial loss of green in the pattern of figure 6. In this leaf the loss appeared in one side of the leaf. Painted on December 4, 1913.

23. Leaf classed as *green-yellow spotted-red blotched*. The red epidermal blotches are large and much coalesced, which is a frequent variation from the condition of figure 5.

24 and 24a. Lower and upper surfaces of the same leaf from a young plant grown from a bud variation which involved a sectorial loss of red, giving *solid red upper center* from *solid red* on both surfaces. In this leaf these two types are sharply limited to one half of the leaf. The young plant exhibited loss of yellow by fluctuating variation. The leaf painted was an upper leaf showing no yellow, but at the same time the older basal leaves possessed much yellow, as in type of figure 10.

25. Leaf painted on March 25, 1913, showing sudden bud variation affecting only the half of a leaf. The plant possessed leaves of the type figure 2 shown in the left side. A reversal of the relative position of the green and the yellow gave the pattern of figure 6 in the right side of the leaf, as shown.

26. Taken on March 28, 1913, showing the fluctuational decrease of yellow, giving a poorly defined yellow border. During the following summer the plant returned to the typical form of type 2. This leaf is quite typical of the increase of green during winter seen in numerous plants of type 2, as mentioned on page 22 of the text.

27. Painted on July 27, 1913, showing an irregular pattern with yellow at the border about the apex of the leaf and illustrating a fluctuation produced from both types 2 and 4. Leaf also shows irregular distribution of epidermal red.

28. Leaf classed as *green-red blotched*, but with few large blotches somewhat run together.

29. Leaf also classed as *green-red blotched*, but with fine blotches much coalesced.

PLATE 4.

Three plants photographed on January 16, 1914, all descended from branch 2 of plant 1.

Plant 125111 has leaves uniformly entire and of the pattern *green-yellow-red blotched* (fig. 2). Tips pinched off to prevent early flowering in greenhouse.

Plant 1251412 has leaves uniformly entire and of the pattern *green-red blotched*.

Plant 123153. Old leaves entire, youngest leaves deeply laciniate. Typical condition during early winter for plants showing the fluctuation in leaf shape.

CPSIA information can be obtained
at www.ICGtesting.com
Printed in the USA
BVHW090822070119
537207BV00021B/2073/P

9 781331 955078